New Generation of Organic Fertilizers

Edited by Metin Turan and Ertan Yildirim

Published in London, United Kingdom

IntechOpen

Supporting open minds since 2005

New Generation of Organic Fertilizers
http://dx.doi.org/10.5772/intechopen.95683
Edited by Metin Turan and Ertan Yildirim

Contributors
Shaima Mohamed Nabil Moustafa, Rana H. Taha, Chaichi Devi, Meena Khwairakpam, Mavis Badu Brempong, Abigail Addo-Danso, Daniel Murta, Regina Menino, Alabi Olusoji David, Sanjeevakumar M Hatture, Pallavi V Yankati, Rashmi Saini, Rashmi P Karchi, Helen Avery, Milie Lionelle Tsouga Manga, René Menoh A Ngon, Etienne Akoutou Mvondo, Eunice Ndo, Lucien Bidzanga Nomo, Zachée Ambang, Sonia Sena Alfaia, Eleano Rodrigues Da Silva, Marta Iria da Costa Ayres, Acácia Lima Neves, Katell Uguen, Luiz Antônio Oliveira

First published in London, United Kingdom, 2022 by IntechOpen
IntechOpen is the global imprint of INTECHOPEN LIMITED, registered in England and Wales, registration number: 11086078, 5 Princes Gate Court, London, SW7 2QJ, United Kingdom
Printed in Croatia

British Library Cataloguing-in-Publication Data
A catalogue record for this book is available from the British Library

Additional hard and PDF copies can be obtained from orders@intechopen.com

New Generation of Organic Fertilizers
Edited by Metin Turan and Ertan Yildirim
p. cm.
Print ISBN 978-1-83969-212-3
Online ISBN 978-1-83969-213-0
eBook (PDF) ISBN 978-1-83969-938-2

For EU product safety concerns:
IN TECH d.o.o., Prolaz Marije Krucifikse Kozulić 3, 51000 Rijeka, Croatia, info@intechopen.com or visit our website at intechopen.com.

Meet the editors

Metin Turan received his Ph.D. from the Department of Soil Science and Plant Nutrition, Atatürk University, Turkey in 2002. He is currently a professor in the Department of Genetics and Bioengineering, Yeditepe University, Turkey. His research mainly focuses on soil ecology and biological fertilizer applications. Dr. Turan has more than 100 research publications to his credit. He is a member of many international organizations such as the International Federation of Organic Agriculture Movements (IFOAM), Research Institute of Organic Agriculture (FiBL), and European Biostimulants Industry Council (EBIC), and he has chaired many conferences in Turkey and Europe.

Ertan Yildirim is a full professor in the Department of Horticulture, Atatürk University, Turkey. He received his Ph.D. in Horticulture from the same university in 2003. His research focuses on vegetable growing, vegetable breeding, greenhouse management, seed germination and physiology, organic agriculture, and stress physiology. Dr. Yildirim has more than 200 publications to his credit.

Contents

Preface

Both developing and underdeveloped countries have adopted strategies to produce economically optimal agricultural products to feed the world's growing population, which is expected to reach 9 billion in the first half of this century according to the United Nations Food and Agriculture Organization (FAO). Since the main goal of these strategies was to increase the amount of production, soil productivity was prioritized, whereas the fertility factor of soil was ignored. To increase productivity, chemical fertilizers were used excessively and uncontrollably, which caused many short- and long-term problems in the fertility of the soil, continuity of biodiversity, and the sustainability of the balance in nature. Improper fertilizer applications led to an increase in soil salinity, insufficiency of nutrient elements in soil, increased diseases and pests, pollution of the environment, agricultural lands, and drinking water and rivers, and contributed to climate change, which increased the biotic and abiotic stresses on plants and harmed sustainability. As a result, the use of chemical fertilizers decreased agricultural production in the long term. This outcome triggered the efforts to develop agricultural activities involving organic inputs to regain biodiversity, maintain the ecological balance, and decrease the adverse effects of chemicals on human health.

By definition, a fertilizer is any material of natural or synthetic origin that is applied to soil or plant tissues to provide at least one, but often more, of the nutrients required for plant growth. Most fertilizers that are currently used in commercial agriculture provide the three main nutrients: nitrogen, phosphorus, and potassium. These fertilizers are produced industrially from petroleum and petroleum-based products. On the contrary, organic fertilizers are derived from natural sources such as vegetable matter (e.g., compost and crop residues), animal manure, natural rocks, algae extracts, and so on. They are mostly richer in composition compared to chemical fertilizers and provide more nutritional elements as well as beneficial microorganisms. One of the main advantages of using organic fertilizers is that they increase the resistance of plants against biotic and abiotic stress conditions by boosting their immune systems. Moreover, they provide a slow, long, and sustained release of nutritional elements for plant uptake.

Several natural resources, biological products (beneficial bacteria, lichens, algae, and fungi), mineralization products of plant and animal residues (leonardite, humic acid, rock phosphate, bat manure), fermentation products of waste by worms (worm manure), and products obtained from waste of slaughterhouses (amino acid products) are used in agricultural production as organic fertilizers. Although the use of these environmentally friendly products enables the regeneration of natural resources, improves the productivity and fertility parameters of soil, and contributes to carbon footprint management and sustainability, the substitution of chemical fertilizers with these organic products by producers is still slow due to lack of awareness. These products will get greater recognition only when their benefits are explained thoroughly and scientifically.

Besides the aforementioned natural sources, solid wastes of organic origin in the agricultural and industrial wastes can be converted into new organic products to be used in agriculture or into renewable energy by biogas production technologies. Despite all the benefits, the industrial revolution has also brought an important environmental problem: significant amounts of waste that need to be discarded. The utilization of billion cubic meters/year of agricultural and industrial waste as energy or as organic fertilizer is an important value-added solution to this environmental problem and a profound contribution to sustainability.

Today, sustainable use of agricultural soils is no longer a preference, but rather is a mandatory requirement due to the abiotic adverse conditions such as drought and soil degradation caused by global climate change. For this reason, incorporation of organic fertilizers into the soil is essential to develop the cluster structure (aggregate) that holds the soil particles together, increases the amount of organic matter in the soil, regulates the soil pH, and decreases salinity to increase the water-retention capacity as well as the water and air order of the soil. The use of organic fertilizers ensures both the productivity/fertility of the soil and the sustainable production of healthy food products to feed the increasing world population. Therefore, there is a dire need to increase the awareness of each and every stakeholder in the agricultural industry about the benefits of organic inputs and to rapidly include these inputs in the existing agricultural models, which will provide maximum efficiency with minimum or no negative impact on the environment and increase yield per unit of land. The selection of proper fertilizers and their rational use are the key factors to overcoming the challenge of establishing sustainable economic production. Not only is the type of organic fertilizers used important for a successful outcome, but so are the application method, time, and dosage. It is important that national administrators and suppliers promote the use of organic inputs, educate producers, and ensure the availability of organic products in the market.

This book provides an update on the sources, production, and applications of organic fertilizers and highlights their importance in terms of sustainable agriculture, biodiversity, and the environment. It presents new approaches, ideas, and trends to scientists and producers in this field on how to (1) increase the resistance of plants against biotic and abiotic stress conditions and (2) increase the effectiveness of chemical fertilizers while decreasing the amounts used by using organic inputs that are composed of hormones, antioxidants, enzymes, and amino acids.

Metin Turan
Yeditepe University,
Istanbul, Turkey

Ertan Yildirim
Atatürk University,
Erzurum, Turkey

Chapter 1

Improving Soil Fertility with Organic Fertilizers

Mavis Badu Brempong and Abigail Addo-Danso

Abstract

Organic fertilizers with low C:N ratios can be applied to supply both macro and micronutrients to the soil. Aside nutrient supply, they can improve soil structure, texture, water holding capacity and nutrient holding capacity. The mechanisms that may interplay to allow organic fertilizers to affect the soil and crop yields may include improved nutrient synchrony, general improvement in fertility and/or priming effects. The rate, timing and method of organic fertilizer application must be considered to reduce N and P losses during organic fertilizer application. To meet the nutrient requirement of crops, organic fertilizers must be applied in large quantities, so it is more prudent to apply locally available resources. In a case study where sole organic fertilizer, sole inorganic fertilizer and their combinations were applied under rain-fed conditions, it was observed that manure had the potential to hold nutrients longer. This is a positive finding for drought prone areas.

Keywords: climate smart agriculture, manure, water holding capacity, nutrient holding capacity

1. Introduction

Poor soil fertility is the major biophysical factor affecting crop production in the world [1]. It is a major threat to food security considering the ever-increasing growth rate in human population which is projected to reach about 10 billion by 2050 [2]. In times of old, forests and marginal lands were converted to farmlands to meet the food demands of the growing population. This practice caused the extinction and endangerment of many plant and animal species; hence it is frowned upon by many stakeholders. As such, today, it would not be prudent to encroach land reserves and other marginal lands for agricultural purposes. It is therefore imperative that we improve soil fertility and health of the available land, to increase food production and to ensure the world's food security under the current and projected climate change.

Previously, the use of mineral fertilizer was thought of as the most appropriate remedy to soil fertility problems due to its rapid nutrient release [3]. However, mineral fertilizer lacks the ability to improve the soil's physical properties causing fertility improvement by fertilizers alone to be unsustainable. Over-reliance on mineral fertilizer without due diligence to the organics may lead to increased soil erosion, surface and groundwater contamination, increased greenhouse gas emission and reduced biodiversity [4]. In addition, mineral fertilizers are expensive, and many farmers may not have the purchasing power to acquire it [5]. As a result, the attention of various stakeholders has been drawn to use of organic resources [6].

The application of organic fertilizers presents a more sustainable method of food production. There is unending literature reporting the efficiency and effectiveness of organic nutrient sources in maintaining soil quality (physical, biological and chemical properties), improving crop yields and sustaining productivity [6–8]. The benefits of applying organic fertilizers to the soil are elaborated in this chapter.

2. Types of organic fertilizers

Organic nutrient sources are specifically derived from plant and animal origins [9]. They include plant residues, animal wastes and biofertilizers. In this era where climate change and the COVID-19 pandemic has impacted agricultural production and the financial capabilities of all workforces including farmers, farmers could use organic fertilizers available to them for soil fertility purposes because they are cheaper and more environmentally friendly when they are locally available [10].

Organic fertilizers include poultry manure, cattle manure, green manure (often legumes), field crop residues, composts, bone meal, household waste, blood meal, slurry, cocoa pod husks, palm kernel cake, among others. Biofertilizers are products containing single micro-organisms or combinations of them which when applied help fix atmospheric N, solubilize nutrients, mobilize nutrients, or secrete growth promoting substances to aid crop growth. These products do not supply nutrients themselves but enhance the activities of soil microbes to make more nutrients available to crops. They are categorized into N-fixing biofertilizers, phosphorus solubilizing biofertilizers, composting accelerators and plant growth promoting rhizobacteria [9]. Most of the plant and animal residues are often by-products and nuisance to the environment. Using them as nutrient sources would help reduce waste and greenhouse gas emission.

3. Benefits of organic fertilizers to the soil

3.1 Balanced nutrient supply

Organic fertilizers supply all essential crop nutrients (N, P, K, S, Ca, Mg, B, Cl, Cu, Fe, Mn, Mo, Ni and Zn) in balanced forms, including micronutrients. This is often not the case for any one inorganic fertilizer. Since all these nutrients make up the biomass of organic residues, they are released during the decomposition process into the soil. The downside to applying organic fertilizers alone is that they contain very minimal amounts of these nutrients and as such must be applied in bulky quantities to meet crop nutrient demands [11]. Also, the fact that only a fraction of the nutrients in organic fertilizers can be released per season must be factored in when applying organic fertilizers. On the average, as a rule of thumb, only about 50% of nutrients in organic fertilizers are mineralized in the first season of application [12]. Usually, the focal nutrient used to calculate the amount of organic fertilizer to apply is its nitrogen (N) concentration. For example, 30% decomposed cattle manure (DCM) contains about 2% N [13]. Assuming a farmer grows maize, which requires about 90 kg/ha N, that means:

$$100\,\text{kg DCM} = 2\,\text{kg N}. \tag{1}$$

Therefore, to supply 90 kg N = (90 × 100)/2 = 4500 kg DCM.

Since the applied DCM will only supply half the amount of N required in a season, the amount must be doubled to make **9000 kg** DCM, thus **9 t** of DCM. [It should be noted that the C:N ratio of the organic material must be below 25 [14] to ensure that all decomposition and mineralization requirements are met. The organic material should also be tested for nutrient concentrations or at least the focal nutrient concentration to use as basis for the calculation].

To supply same amount of N through mineral fertilizer, a farmer would only need about 200 kg Urea, however, in organic applications, other nutrients are concurrently being applied. Since a large amount of DCM would supply the required N and other nutrients, it must be available to the farmer. Hence advocates of organic fertilizers must emphasize on ways to raise such large amounts of materials for application locally if sole organic production is desired.

3.2 Improving nutrient holding capacity

Aside the balanced nutrient supply, organic fertilizers add organic matter to the soil if a long-term application is practiced. Organic matter improves the nutrient holding capacity of the soil because it contains organic acids that increase the H^+ ions and surface charge of the soil, causing the soil's cation exchange capacity to increase [15]. Thus, the soil's ability to hold more cations (nutrients) at exchange sites is increased and hence the nutrient holding capacity of the soil is also improved. Organic matter also improves the buffer capacity of the soil and increases the soil's ability to resist a change in pH, which in turn affects nutrient loss or gain to the soil [16]. Organic fertilizers increase microbial activity in the soil, causing increased nutrient mineralization rates for the benefit of crops. They stimulate the activities of aerobic and anaerobic bacteria [17] and arbuscular mycorrhizae fungi that form networks of root extension for extensive nutrient availability to crops. Upon the lysis and decomposition of soil microbes, nutrients retained in their biomass are made available in the soil and to crops.

3.3 Improving water holding capacity

Soil structure, texture, bulk density, and organic matter content are the controls on soil water holding capacity; therefore, any management practice that improves these soil properties, in turn, improves water holding capacity (WHC) of the soil. Soil moisture content is largely dependent on the specific surface area of the soil and the thickness of films of water surrounding the pores [18]. The addition of organic matter through organic fertilizer application improves soil aggregation and increases the surface area of the soil, presenting the soil with more room for soil particles to be surrounded by films of water. As a result, the soil can hold more water against the pull of gravity which drains water from the soil.

While soil organic matter binds soil particles, it also stimulates the activity of soil microfauna whose movement create micro and macropores in the soil, creating extra room for water infiltration [19]. Thus, soil water holding capacity can be improved by the addition of organic fertilizers. In the wake of climate change, where unexpected droughts may be imminent, improving the water holding capacities of the soil with the application of organic fertilizer is the way to go. Also, the physical presence of organic materials on the soil serves as mulch that reduces evaporation and retains moisture in the soil. It also reduces the speed of runoff water and allows rain or irrigation water to infiltrate the soil at favorable speed, thereby reducing erosion, soil and nutrient loss [19].

3.4 Improving soil texture and structure

The soil binding properties of organic matter and improvement in soil aggregation helps to improve soil structure [20]. The addition of organic matter also improves soil texture and aeration. Soils with improved structure and texture allow easy air, water, and root movement to support healthy crop growth.

4. Mechanisms underlying organic fertilizer effects on soil

Many research works have observed extra crop yields with organic fertilizer application compared to when its nutrient equivalents are applied through mineral fertilizer [21–23]. Various mechanisms have been proposed to explain this added crop yields from organic fertilizer application. Some of which include **improved nutrient synchrony, priming effect,** and **general fertility improvement**.

Under the improved nutrient synchrony mechanism proposed by Vanlauwe et al. [23], when organic fertilizers are applied, they supply microbes with energy from the carbon they contain, to drive decomposition processes. This leads to temporal immobilization of soil N [24, 25] to build their body tissues. The immobilized N is made available at a later stage of plant growth when the microbes have decomposed the organic material to make nutrients available and/or some microbes have lysed and released their nutrients to the plant when it needs nutrients most. In effect, the peak of nutrient supply coincides with highest crop nutrient demand point when crops have matured, so that the nutrients are efficiently utilized, and little is lost to the environment. Kapkiyai et al. [26] reported that a combination of organic and mineral nutrient sources has been shown to result into synergy and improved synchronization of nutrient release and nutrient demand and uptake by plants leading to higher yields.

The general fertility improvement mechanism [23] is based on the theory that organic matter, aside its addition of nutrients to the soil, improves other physical properties of the soil that helps to perpetuate the nutrient addition effect in real time. Some of these benefits include the improvement of soil structure, water and nutrient holding capacities as discussed above. It also adds micronutrients which is usually not the focus of inorganic fertilizer application.

Priming effect is another mechanism proposed by Kuzyakov et al. [27], in which organic fertilizers affect additional crop yields. Priming refers to strong short - term changes in the turnover of soil nutrients caused by the addition of easily decomposable organic materials. Changes may be positive or negative depending on whether nutrients are rapidly mineralized or immobilized. Under this mechanism, a sum of nutrients available in the soil after harvest and nutrients in crops from the field are higher than a sum of the initial soil nutrients and nutrients in the organic materials. Thus, the additional unaccountable nutrient is the result of organic fertilizer precursing a more rapid mineralization rate and dissolution of previously unavailable/ fixed nutrients into solution. This is made effective by the improvement in microbial population, diversity and activity affected by the organic material addition.

These mechanisms, though proposed by different authors, all point to the fact that organic fertilizers are beneficial to the soil and consequently, crops.

Despite the benefits of organic fertilizers to the soil, organic resources application is limited by the large amounts required to meet nutrient demand [28]. Hence locally available organic resources must be used to overcome this limitation. In areas where animal production is common, feedlot manure is the most available organic fertilizer resource. Crop residue retention and cash crop- cover crops rotation is an option to increase on-farm residue production.

One other option that has proven to be effective is an integrated nutrient management approach where organic and inorganic fertilizers are applied in right quantities [29]. This approach helps to harness the mechanisms underlying the effects of organic fertilizer application on crops, resulting in synergy in terms of crop yields.

5. Role of organic fertilizer in climate smart agriculture

In times when climate change is imminent and its effect on agriculture tends to endanger food security, it is paramount that farmers and other stakeholders use strategies and resources that adapt farming systems to the changing climate. Climate change is mainly driven by natural and anthropogenic activities that pump greenhouse gases (examples CO_2, CH_4, N_2O) into the atmosphere [30, 31]. It may lead to extreme droughts or extreme floods, which may have devastating impacts on food production and agriculture. In this light, organic fertilizers are a great resort due to their replenishing effects on soil physical and chemical properties. Aside the benefits of organic fertilizers discussed above which may adapt the soil to drought conditions, soils should be well drained and loose in flood prone areas in wait of climate change. In compact and poorly drained soils, the addition of organic fertilizers would improve soil particle aggregation and structure to give the soil more room to infiltrate water without settling on the top for too long to cause floods. The addition of organic matter reduces the inventory of greenhouse gasses contributed to climate change by agriculture. This is achieved by the sequestration of carbon into the soil from organic fertilizers applied. The carbon would have been lost to the atmosphere as CO_2 or CH_4 if it had not been incorporated into the soil [32]. As a result, the application of organic fertilizers to the soil helps to reduce greenhouse gas emission leading to global warming and a consequent climate change and helps adapt the soil to the current and future changes in the climate.

6. Qualities of a good organic fertilizer

Since organic materials are diverse in type and nutrient composition, it is difficult to give a general recommendation of an organic material. The lignin, polyphenol and nitrogen contents of organic material are important controls on its nutrient mineralization, once applied. It is important to evaluate the carbon to nitrogen (C:N) ratio of an organic material to determine if application of the material will lead to N mineralization or immobilization. A C:N ratio of 25 would enhance decomposition and mineralization by soil microbes while a C:N ratio above that would enhance N immobilization [14]. Hence the lower the C:N ratio, the more rapidly nutrients will be made available to the soil. Organic materials high in lignin (>15%) and polyphenol (>5%) contents usually have high C:N ratios and are resistant to microbial decomposition; hence will decompose slowly. If the N content of the material is 2.5% or more, it would likely decompose and mineralize faster [33].

7. How to apply organic fertilizers to harness all the benefit

The effectiveness of an organic material as a fertilizer is also dependent on how it is applied. Surface application of organic fertilizer enhances the loss of N through ammonia volatilization or loss of N and P through runoff and erosion. Judicious methods by which organic materials may be applied to reduce wastage and nutrient losses include band spreading, trailing hose method, burying method, rapid soil incorporation, and the addition of nitrogen inhibitors [34].

Band spreading is the application of the organic material(s) in narrow bands usually a few centimeters away from the crops. This reduces the surface area of the material to the atmosphere so that ammonia volatilization is reduced. To reduce the rate of denitrification as well, band spreading should be done during cool weather with no excessive soil moisture and at right rates. The crop canopies also serve as a physical barrier that further reduces the rate of ammonia volatilization from band spreading applications.

Slurries or liquefied organic fertilizers could be applied in these narrow bands through trailing hoses which hang down from a boom and run along or just above the surface of the soil.

Organic amendments could also be buried at about 5-30 cm depth depending on the crop establishment. Deeper depth burying can be practiced before crops are grown while shallower depth is suited for already established crop fields. This method greatly reduced N loss through ammonia volatilization and the loss of material through erosion.

Manure could be rapidly incorporated into the soil during soil tillage (before planting) or with hand implements to reduce N and P losses in volatilization and runoff.

Under conditions with high denitrification potential, nitrification inhibitors could be added to organic fertilizers to delay the rate at which ammonium is converted to nitrates, which is a suitable substrate that precursors the denitrification process. It is important to apply organic fertilizers at cool times of the day and at the right rates to reduce nutrient losses.

8. Case study (Research)

The sole application of organic fertilizers has proved to be a slow means of nutrient supply to the soil. Hence the combined use of organic and inorganic nutrient sources has been proposed [29]. Such applications harness the benefits of synergistic interaction between the organic and inorganic nutrient sources. The main objective of this research was to increase maize yield with the application of organic manure or a combination of it with mineral fertilizer. To arrive at this objective, the yield of maize following varying rates of combined manure and mineral fertilizer applications were estimated at harvest, synergistic benefits of combined applications were quantified and the effect on soil nutrient stocks were analyzed.

8.1 Methodology

A field experiment was conducted at the plantation section of the Kwame Nkrumah University of Science and Technology under rain-fed conditions. Nine treatments (three levels of mineral fertilizer at 0, 50 and 100% of the 90-60-70 kg/ha NPK recommended rate (RR) by three levels of manure at 0, 50, 100% of 5 t/ha RR) were applied on the field in a factorial fashion arranged in Randomized Complete Block Design (RCBD) with three replications. The land was slashed and burned and later plowed and harrowed to a fine tilt. Plot layouts were done with lines and pegs with each plot measuring 3 m by 2 m. There were 2 m alleys between replications and 1 m alleys between plots. Initial soil and manure sampling and analyses were done to characterize them. Randomized manure treatments units were allocated to their designated plots. The Akposoe maize variety developed by the Crops Research Institute of Ghana was planted 2 weeks after manure allocation.

Weeding was done manually when necessary. Mineral fertilizer application was done 2 weeks after planting (WAP). The fertilizers were applied as urea, triple superphosphate and murate of potash. The urea was split applied in the first fertilizer application (2 WAP). The other half of the urea was applied 6 WAP. The manure was spread in the plots and raked in to about 5 cm depth. The fertilizer was applied by the band placement method, about 5 cm away from the maize plants. A final soil analysis was done after harvest to determine soil nitrogen (N), phosphorus (P) and potassium (K) levels. Data was subjected to analysis of variance (ANOVA) with the GENSTAT statistical package and significant means were separated with least significant difference at 5%.

Note: Rains were quite erratic at the start of the experiment until an unexpected shortage during the reproductive stage of maize growth. Though unfortunate, this was a good situation to determine if manure applied to the soil would help maintain more soil moisture and consequently impact maize yield.

8.2 Results and discussion

The lack of rains crippled any effect of the manure alone or its combinations with mineral fertilizer to create differences in the yield of maize. Limited soil moisture has been reported to constraint maize yield [35], because all the processes involved in nutrient movement to roots, uptake by roots and translocation through the transpiration stream use water [36].

After harvest, soil and statistical analysis showed that plots receiving 50 and 100% rates of manure had a significant 20% more total soil N than the control and mineral fertilizer rates. It is possible that due to the rapid nutrient release mechanism of mineral fertilizer, most of its nutrients was released during the early stages of the maize growth, subject to rapid loss from the soil system. The C:N ratio of the manure was 23.08, which is an indication that N was being mineralized [35] into the soil system over a long period, even after the shortage of rains. A combined use of the full rate of manure and full rate of mineral fertilizer also had 20% more total soil N than each individual nutrient source. It is evident that combining organic and inorganic inputs creates a balance between increasing N availability for plant uptake over sole organic application and decreasing N availability for potential system losses compared to fertilizer alone [37].

The rather erratic rains at the beginning of the experiment might have caused soil P and K to leach beyond root zone, hence the lack of differences between the effects of sole manure and mineral fertilizer applications or their combinations at the end of the experiment.

8.3 Conclusion

Overall, it was concluded that organic manure had the potential to hold nutrients in the soil longer than inorganic fertilizers. In the advent of climate change, it could be a very useful tool especially in areas was droughts are expected.

Author details

Mavis Badu Brempong* and Abigail Addo-Danso
Crop Research Institute, Kumasi, Ghana

*Address all correspondence to: abulgo@yahoo.co.uk

References

[1] Tandzi NL, Mutengwa SC. Factors affecting yield of crops. Agronomy SP—Ch. 2 UR—DO—10.5772/intechopen.90672 SN—978-1-83881-223-2 PB—IntechOpen CY—Rijeka Y2-2021-07-02 ER—. 2019. DOI: 10.5772/intechopen.90672

[2] Bruinsma J. The resource outlook to 2050. By how much do land, water and crop yields need to increase by 2050? In: FAO Expert Meeting on How to Feed the World in 2050. Rome, Italy: FAO; 2009. p. 33

[3] International Food Policy Research Institute (IFPRI). Sustainable Options for Ending Hunger and Poverty. Green Revolution: Blessing or Curse. Washington DC, USA; 2002

[4] Tilman D, Cassman KG, Matson PA, Naylor R, Polasky S. Agricultural sustainability and intensive production practices. Nature. 2002;**418**:671-677

[5] Tetteh FM, Issaka RN, Ennin S, Buri MM. Soil Fertility Initiative, Fertilizer Update and Reccommendation Trials. Ghana: Soil Research Institute; 2008. p. 33

[6] Suge JK, Omunyin ME, Omami EN. Effect of organic and inorganic sources of fertilizer on growth, yield and fruit quality of eggplant (*Solanum melongena* L). Archives of Applied Science Research. 2011;**3**:470-479

[7] Negassa W, Negisho K, Friesen DK, Ransom J, Yadessa A. Determination of optimum farmyard manure and N, P fertilizers for maize on farmers' fields. In: Seventh Eastern and Southern Africa Regional Maize Conference, 11th–15th February 2001. 2001. pp. 387-393

[8] Vanlauwe B, Diels J, Aihou K, Iwuafor ENO, Lyasse O, Sanginga N, et al. Direct interactions between N fertilizer and organic matter: Evidence from trials with 15 N-labelled fertilizer. In: Vanlauwe B, Diels J, Sanginga N, Merckx R, editors. Integrated Plant Nutrient Management in Sub-Saharan Africa. Wallingford: CAB International; 2002. pp. 173-184

[9] FAO. Plant nutrition for food security. A guide to integrated nutrient management. In: Roy RN, Finck A, Blair GJ, Tandon HLS. 2006. ISBN: 92-5-105490-8. Available from: https://www.fao.org/3/a0443e/a0443e.pdf

[10] Adejobi KB, Fameye AO, Akenbi OSO, Adeosun SA, Nduka AB, Adeniyi DO. Potentials of cocoa pod husk ash as fertilizer and liming materials on nutrient uptake and growth performance of cocoa. Research Journal of Agriculture and Environmental Management. 2013;**2**(9):243-251

[11] Timsina J. Can organic sources of nutrients increase crop yields to meet global food demand? Review paper. Agronomy. 2018;**8**:214

[12] Eghball B. Nitrogen mineralization from field-applied beef cattle feedlot manure or compost. Soil Science Society of America Journal. 2000;**64**:2024-2030. DOI: 10.2136/sssaj 2000.64620 24x

[13] BARC (Bangladesh Agricultural Research Council). Fertilizer Recommendation Guide 2012. Soils Pub. No. 45. Farmgate, Dhaka: Bangladesh Agricultural Research Council; 2012

[14] Brust GE. Management Strategies for Organic Vegetable Fertility. Safety and Practice for Organic Food2019. pp. 397-408. DOI: 10.1016/b978-0-12-812060-6.00009-x

[15] Havlin JL, Tisdale SL, Nelson WL, Beaton JD. Soil Fertility and Fertilizers. An Introduction to Nutrient Management. 8th ed. Upper Saddle River,

New Jersey: Pearson Educational, Inc.; 2014. p. 98 ISBN 978-81-203-4868-4

[16] Buresh RJ, Dobermann A. Organic materials and rice. In: Annual Rice Forum 2009: Revisiting the Organic Fertilizer Issue in Rice. Vol. 2010. College, Laguna, Phiilippines: Asia Rice Foundation; 2010. pp. 17-33

[17] Mamaril CP, Castillo MB, Sebastian LS. Facts and Myths about Organic Fertilizers. Muñoz, Nueva Ecija, Philippines: Philippine Rice Research Institute (PhilRice); 2019

[18] Vengadaramana A, Thairiyanathan JJP. Effect of organic fertilizers on the water holding capacity of soil in different terrains of Jaffna peninsula in Sri Lanka. The Journal of Natural Product and Plant Resources. 2012;**2**(4):500-503

[19] FAO. The importance of soil organic matter. Key to drought-resistant soil and sustained food production. FAO Soils Bulletin 80. 2005.

[20] Funderburg, E. What Does Organic Matter Do in Soil?. 2001. Available from: https://www.noble.org/news/publications/ag-news-and-views/2001/august/what-does- organic-matter-do-in-soil/

[21] Giller KE. Targetting management of organic resources and mineral fertilizes: Can we match scientists fantasies with farmers' realities? In: Vanlauwe B, Sanginga N, Diels J, Merckx R, editors. Balanced Nutrient Management Systems for the Moist Savannah and Humid Forest Zones of Africa. Wallingford, U.K.: CAB International; 2002. pp. 155-177

[22] Palm CA, Myers RJK, Nandwa SM. Organic-inorganic nutrient interaction in soil fertility replenishment. In: Buresh RJ, Sanchez PA, Calhoun F, editors. Replenishing Soil Fertility in Africa. Madison, Wisconsin: Soil Science society of America; 1997. pp. 193-218

[23] Vanlauwe B, Wendt J, Diels J. Combined application of organic matter and fertilizer. In: Tian G, Ishida F, Keating JDH, editors. Sustaining Soil Fertility in West Africa SSSA Special Publication no. 58. Vol. 9 (Jan–Jun 2006). Madison, USA: Soil Science Society of America; 2001. pp. 247, 12-280, 18

[24] Myers MG, McGarity JW. The urease activity in profiles of five great soil groups from northern New South Wales. Plant and Soil. 1968;**28**(1):25-37

[25] Palm CA, Gachengo CN, Delve RJ, Cadisch G, Giller KE. Organic inputs for soil fertility management: Some rules and tools. Agriculture Ecosystems and Environment. 2001;**83**:27-42

[26] Kapkiyai JJ, Karanja NK, Woomer PL, Qureshi JN. Soil organic carbon fractions in a long-term experiment and the potential for their use as a diagnostic assays in highland farming's of Central Kenya. African Crop Science Journal. 1998;**6**:19-28

[27] Kuzyakov Y, Friedelb JK, Stahra K. Review of mechanisms and quantification of priming effects. Journal of Soil Biology and Biochemistry. Institute of Soil Science and Land Evaluation, University of Hohenheim, D-70593 Stuttgart, Germany. Institute of Organic Farming, University of Agricultural Sciences, A-1180 Vienna, Austria; 2000;**32**:1485-1498

[28] Vanlauwe B, Giller KE. Popular myths around soil fertility management in sub-Saharan Africa. Agriculture, Ecosystems and Environment. 2006;**116**(1-2):34-46. DOI: 10.1016/j.agee.2006.03.016

[29] Brempong MB, Opoku A, Ewusi-Mensah N, Abaidoo RC.

Evaluating added benefits from combined cattle manure and fertilizer application in am maize cropping system. Journal of Environmental Science and Engineering. 2017;**B6**:34-40

[30] Denchak M. Global Climate Change: What You Need to Know. 2017. Available from: https://www.nrdc.org/stories/global-climate-change-what-you-need-know

[31] Nwankwoala HNL. Causes of climate and environmental changes: The need for environmental-friendly education policy in Nigeria. Journal of Education and Practice. 2015;**6**:30

[32] Slave C, Man C. The contribution of human activities to climate changes. In: Agrarian Economy and Rural Development—Realities and Perspectives for Romania. 3rd Edition of the International Symposium, October 2012, Bucharest. Bucharest: The Research Institute for Agricultural Economy and Rural Development (ICEADR); 2012. pp. 292-295

[33] Haile W, Tesfamaraiam AA. Potential of local plants as a source of NPK on small holder fields in Southern Ethiopia. Technical Report. ISBN: 978-9988-633-73-82013. pp. 8-10

[34] Misselbrook T, Bittman S, Cordovil CMdS, Rees B, Sylvester-Bradley R, Olesen J, Vallejo A. Field Application of Organic and Inorganic Fertilizers and Manure. Draft Section for a Guidance Document. 2019

[35] Bationo A, Waswa B, Okeyo MJ, Maina F, Kihara J. Innovations as key to the gree revolution in Africa. Exploring the Scientific facts. 2007;**1**:123-133

[36] Olsen SR, Kemper WD. Movement of nutrients to plant roots. Advances in Agronomy. 1968;**20**:91-149

[37] Gentile R, Vanlauwe B, Chivenge P. Interactive effects from combining fertilizer and organic residue inputs on nitrogen transformations. Journal of Soil Biology and Biochemistry. 2008;**40**:2375-2384

Chapter 2

Restoration of Soil Organic Carbon a Reliable Sustenance for a Healthy Ecosystem

Alabi Olusoji David

Abstract

Agricultural sustainability is an indicator for economic prospect across the globe. The revolution of industrial development and the growth of annual crop to meet the need of increasing world population is a determining factor for SOC availability. Sustainability of agriculture is largely related to SOC and management practices. Agro-ecological stability is significant to soil type and fertility input. Organic matter is a combination of plant residue and/ or animal waste. This is capable of accumulating carbon and nitrogen in the soil. It retains water and support the buildup of organic carbon. It enhances the stability of SOC and crop yield. The use of organic matter is effective at stabilizing the microbial communities. Carbon sequestration is high with crops that have abundant residues. SOC can potentially mitigate climate change. It prevents the use of minimum and conventional tillage. Higher deposit of SOC is associated with crop yield. Perennial crop cultivation such as cup plant (*Siliphium perforliatum. L.*) can potentially sequestrate carbon into the soil than annual crop. SOC are often exhausted with the cultivation of annual crop such as maize. However, SOC can be retained by growing clover in between harvests and the next sowing. Mineral fertilizer can likewise accumulate SOC but not as efficient as the use organic manure and plant residue. Perennial crop was found useful at preventing environmental degradation and soil compaction. Consistent assessment of SOC is essential for continuous food production and plant growth. This can be achieved through a multidimensional software called multiple linear regression.

Keywords: soil organic carbon, sustenance, ecosystem

1. Introduction

Global warming is caused by the continuous increase in greenhouse gases in the earth's atmospheric surface [1, 2]. 50% of the total global emission is from agricultural production [3]. Excessive application of inorganic fertilizer causes ammonia volatilization, soil nitrogen leaching, air pollution, and soil acidification [1, 4]. This is a result of a rapid increase in the global population. They suggest the need for the protection of soil quality to meet the food requirement of the increasing global population and societal development [5, 6]. Climate change significantly affects the atmospheric carbon pool and its availability in the soil for agricultural productivity [7]. The loss of soil organic carbon and land degradation is one of the major factors that affect the yield of existing farmland [3, 8]. Soil organic matter deposition is released from approximately two-third of CO_2 exchange between the terrestrial ecosystem and atmosphere [9, 10]. The most active part of the global carbon pool is soil farmland. Farmland soil enables global carbon and nitrogen cycling [4, 11]. Farmland is an important source and sinks of CO_2 for the farmland ecosystem. Agricultural farmland with decrease organic carbon has been proven to be unreliable [6, 12]. This has led to pressure in crop production through cropland management practices such as irrigation and fertilization [8, 13]. Organic matter plays a significant role in food production and agricultural land expansion. It is an environmental factor that determines sustainability and development [8, 13]. It is an indicator for factors such as water retention and nutrient availability [14]. It is a structural balance that promotes efficient drainage, aeration and minimizes loss of topsoil from erosion [6, 15]. It decreases reliance on external inputs such as fertilizer and irrigation [11, 16]. It is a stable and a last longing input that supports crop yield and sustainability [11, 16]. One of the major limitations to the availability of soil organic matter in the soil is the target method of prediction for a specific agricultural land and environmental development [17, 18]. Yield is often expected to increase per unit area with a certain measure of organic matter. But yardstick for determining organic matter is yet to be fully exploited [6, 19]. Several studies have been established in relationship with soil organic matter and yield [6, 19]. It was also found that a concrete agreement among researchers is yet to be reached. For instance, a decrease [20], an increase [21], and no change were recorded in some research findings [11]. Variation in result findings may be due to management, climate change, and soil type [11, 20, 21]. Likewise, a global understanding of soil organic with yield is heterogeneous [11, 20, 21]. This suggests the need to test and understand the effect of soil organic matter and carbon in agriculture and its environment [11, 20, 21]. This paperwork investigates the impact of organic carbon and its relevance in agricultural sustainability and development.

2. Carbon oxidation and deposition in the soil

Organic carbon is an important component of the soil that can stimulate functional compounds and enhances the performance of a microbial community [22, 23]. It provides lubrication and facilitates ease of energy transfer in an ecosystem [6, 8]. Cropping history and management on agricultural land are determining factors for net carbon sink or CO_2 availability in the soil [7, 23]. Arable crop production decline carbon stored between 40% - 60% through conventional tillage and planting activities [9, 24]. Though at some point depending on the soil type and climate change crop production may attain an equilibrium carbon [9, 24]. This attainment may vary with different agro-ecological zones or regions [25, 26]. Soil carbon accumulation largely depends on the rate at which biomass decomposes [12, 22]. This includes primary reduction of plant residues above ground; leaves, stems, and other tissues. Likewise, plant exudates and the below ground as well as the roots. The pool of carbon in the soil depends on the rate at which organic carbon is oxidized by microbes and invertebrates [12, 22]. Soil organic carbon can be categories into active, stable, and inert organic carbon pools; active organic carbon is closely related to the nutrient requirement supply into the soil [7, 27]. They are permeable and can efficiently build up organic matter [13]. This category comprises microbial biomass carbon, soluble organic, mineralisable carbon, and carbohydrate, stable organic carbon pools are dominated by carbohydrate and lipid. This pool is made of organic particles and carbohydrates [15, 22]. The release of nutrients in this case is relatively slow. In other words, the inert organic carbon pool deposition is very slow. It comprises lignin, hummus, polyphenol, and polysaccharide [15, 22]. More carbon will be available in the soil if organic carbon oxidation is low but with high oxidation, soil organic matter is used up than it is replaced by new biomass [2, 5]. A low oxidation rate increases the net sink of CO_2. However, in the soil where biomass deposition is stable, the net change in carbon will be stored is continually [2, 5]. The continuous application of biomass at a point attains maximum carbon saturation. Equilibrium change of CO_2 uptake into the atmosphere plays a significant role in net carbon sink in the soil [2, 5]. The quality of organic matter is affected by deposition. Microclimate and microorganisms play a major role in steady metabolization processes [20, 22, 27]. This regulates the extent to which chemical constituent is released into the soil. It was recorded that senesced leaves and stems with high carbon-nitrogen and corn stalks as well as the wood decomposes at a very low rate [28, 29]. Nevertheless, legumes and stems decompose more rapidly due to their low carbon-nitrogen and lignin content. The aggregate of oxygen supply to microbes in the soil may be restricted by tillage due to accelerated deposition by breaking apart the soil [15, 30]. This promotes rapid oxidation of carbon into the air. Plowing exposes the soil to direct insolation which may support rapid deposition [3, 31]. In this regard, CO_2 emission is remarkably high and profound. This affects the quantity and quality of organic matter in the microenvironment [11, 22]. The net annual flux of soil CO_2 year after year represents a change in organic soil carbon where erosion is put to control [14, 32]. Erosion prevents direct oxidation of carbon into CO_2 due to surface runoff, particles are washed into waterways to bury existing carbon [14, 32]. It slows its deposition due to accumulation in anaerobic sediments. In other words, a carbon sink may be achieved as a result of slow deposition on an eroded site [14, 32]. It was recorded that the sink capacity of eroded soil amounted to 26% in a finding [3, 14, 15]. Several chemicals used in agricultural land increase CO_2 in the air. Its impact on global warming is highly detrimental for instance 2.3 kg CO_2 per active ingredient (750 Ngm^{-1}) 4.5 kg CO_2 per kg of N methane (CH_4) are applied as fossil fuel to meet its temperature and pressure requirement [31, 33]. Also, lime ($CaCO_3$) and dolomite ($CaMg(CO_3)_2$) were applied to neutralize the effect of acid cation in the soil [2, 32]. The use of lime may be influential as an agent of weathering. Nitric acid produces nitrifying bacteria [1, 4]. This supports the breakdown of carbon in lime.

$$CaCO_3 + 2HNO_3 \rightarrow Ca^+ + 2HNO_3^- + H_2O + CO_2$$

But a weak carbonic acid from the root acid and microbial respiration transform lime into bicarbonate [1, 4].

$$CaCO_3 + H_2O + CO_2 \rightarrow Ca + 2HNO_3^-$$

Dissolution of lime by strong acid generates a large amount of CO_2 into the air but weak acid support carbon sinks into the soil [15, 34]. It was revealed in research that sequestration of lime carbon depends on nitric or carbonic acid [24, 35]. CO_2 may largely be neutralized for agricultural lime [3, 36]. The growth of biomass both above and below ground represents CO_2 sink as carbon is captured in perennial vegetative growth [2, 37]. Perennial crops increase carbon sequestration due to persistent land cover [2, 37]. In this case, soil carbon is continually stored in the soil [37]. Perennial crop accumulates more carbon than annual crops [31]. Land cover prolong carbon stored in the soil. However, soil biomass is often accomplished by reducing the soil deposition rate. It increases soil organic carbon [38, 39]. Carbon sequestration is high with crops that have abundant residue [6, 11]. Soil carbon can be enhanced through corn rotation. The use of corn-soybean rotation was found to be effective than corn rotation [12, 16]. The biochemical complexity of carbon residue input in the soil is a result of land cover. This helps build soil carbon [12, 16]. The no-till practice performs the function of deepening carbon into the soil. It increases the storage at the surface layer. Soil carbon content is ample evidence for no-till gain [8, 9]. More so, agricultural CO_2 could be derived from energy input and soil amendment [25, 35]. Most cropping systems are not effective at retaining carbon as exogenous input such as manure, biochar, or sewage sludge [11, 33]. In other words, the native ecosystem is preferred at storing carbon to the use of the organic supplement [2, 25]. It is better concluded that permanent no-till mitigation is by far sustainable at retaining carbon in the soil [20, 40].

3. Climate change and its influence on environmental balance

Climate change is an environmental behavior that characterizes the amount of water availability or dryness in a particular area. It is proportionate to land utility and nutrient availability [24, 25]. A land cover with the forest is significant to accumulating carbon and nitrogen in the soil [16, 27]. In other words, land use and evacuation of forests for buildings are inversely proportional to the balance of greenhouse gas emissions for the atmosphere [24, 26, 35]. This involves activities such as tractor pass on the land, burning of trees, and continuous harvest. This change is substantial to agricultural development and food availability [6, 7, 19]. In recent findings, it was hypothesized that soil depth of 30 cm is rich in carbon than the entire atmosphere. The carbon sink in the ocean is though relatively higher than the amount in the soil [1, 8]. However, soil and forest vegetation play a major role in carbon storage [2, 36, 38]. The global average temperature has risen nearly twice as much as high as the land surface air temperature [34]. Climate change has been noted to harm food security and the terrestrial ecosystem since the pre-industrial revolution [6, 7, 19]. This change was observed between 1850 and 1900 and 2006–2015. The land surface air temperature was found to have increased by 1.53°C and the global mean surface temperature (GMST) by 0.87°C '[34]. This frequent rise in temperature was recorded in the Mediterranean, West Asia, South America, Africa, and North-East Asia [34]. This global change has resulted in vegetation browning than greening in many regions [2, 6, 7]. This is associated with dust storms, evapotranspiration, and decrease precipitation coupled with human activities [14, 28, 29]. This has drastically affected sustainability and development [24, 25]. In the last two decades, it was discovered that continuous rise in average temperature and reduction in the amount of rainfall might create a need for irrigation and poor agricultural output [8, 41]. More so, the success of agricultural output depends on the plant and animal cycle [12, 35]. Global warming may cause a polar shift in the climatic zone in the middle of the equator [1, 2]. The global warmth may result in high latitude. The regions with high latitude suffer drought, wildfire, and pest outbreaks [1, 2]. These regions have been depicted with global warmth of 1.5°C – 3.0°C. This warmth can be ascribed to permafrost degradation, poor agricultural output, and minimal carbon sink [2, 17]. This resulted in soil compaction and dryness [9]. It destabilizes the structural relevance of the soil, poor water retention, and plant growth [24, 40]. The nutritional quality of crops is lowered with increased atmospheric CO_2 [5, 9]. Presently, over 7.6% rise in global crop and economic model of cereal due to climate change predict higher food price, food insecurity, and hunger in 2050 [25, 26]. In this case, food stability is disrupted due to extreme weather conditions [24, 25]. This climate influence on seasonal changes may affect plant blossoming before their pollinators such as insects, birds, etc. are hatched [20, 26]. This may invariably result in flower loss and poor fruit formation [20, 26]. Soil health and agricultural management are hinged on meeting food production for the increasing world population **Figure 1** [16].

Nevertheless, an unprecedented rate of land and freshwater adaptive use for agriculture has been estimated to be 70% [25, 26, 35]. This is a result of global population growth and change in per capita consumption of food, fiber, timbre, and energy release [6, 7, 19]. This global population change has contributed to net GHG emissions, loss of natural ecosystems, and declining biodiversity [2, 7]. Likewise, more than 25 – 30% of food produced is wasted due to climate change [2, 7]. These challenges are aggravated in frost and ice-dominated regions [21, 42]. More so, an area not covered with ice is influenced by human activities [21, 42]. Moreover, due to fossil fuel extinction and greenhouse gas emission, there is an increasing need to replace fossil fuel with biofuel and other plant-based products [40, 41]. Furthermore, erosion is an important determinant of landform and nutrient availability [3, 15]. Intense rainfall, drought, heat loss, heat waves, and a storm cause agricultural land degradation, nutrient loss, plant breakage, stunted growth and total wilting, and perhaps rises in sea level, particularly in the coastal area [3, 43]. Climate change may produce a significant loss in agriculture output by 2050 if measures are not drawn or put in place [25, 26, 35]. Moreover, in some regions of the world climate change is linked to the availability of carbon dioxide and methane in the soil [2, 7]. The permafrost and melt of ice are common in the arboreal region. In this region, an increased temperature causes permafrost to melt [14, 15, 41]. This change over a period trapped organic matter into the frozen which after some time cause a disintegration [3, 43, 44]. Despite the continuous change of climate across the globe, ecosystem and soil quality can be restored by removing carbon dioxide from the atmosphere [1, 38]. According to findings, more than 63 billion tonnes

Figure 1.
Global warming and environmental changes sourced from [14].

of carbon are removed from the soil [1, 38]. The health of the soil is improved by storing carbon underground [2, 37]. The growth of the plant and natural storage of carbon in the soil serves as a defense against climate change [22, 23]. Green space in the cities such as floods and heatwaves is cost-effective protection [25, 26, 35]. They perform the function of flood elimination and storage of excess water [14]. They cool down heat waves due to water accumulation in the soil [14]. They provide a healthy ecosystem during drought through a gradual release of water stored underground [11, 14, 38]. The availability of carbon in the soil can be maintained by converting arable land to grassland [7, 22, 23]. The growth of clover in between harvests and sowing the next crop [8]. This mitigation practice prevents erosion wash. It improves fertility and crop development [12]. Other adaptive measures to fight desertification and land degradation include reduced deforestation, ecosystem conservation, reduced food loss, and waste [3, 15, 30]. The conservation of high carbon ecosystems such as peatland, wetland, rangeland is linked with effective management practice [1, 2]. In addition, carbon sequestration in the soil or vegetation can be maintained by afforestation, reforestation, and agroforestry [12, 29]. The removal of the wood product from the forest restricts carbon in the soil [12, 29]. Peatland is efficient at striking a balance between vegetation and carbon reservoir [22]. This is achieved with the annual removal of CO_2 from the atmosphere when the carbon sink declines towards zero [1]. This is the point where saturation measure is reached between vegetation and carbon reservoir [3, 15]. Management practices such as flood, drought, fire, or pest outbreaks may be hindered by mitigation practice if future management plans are considered [8, 40].

4. Carbon sequestration and sustainable practice

Perennial energy crop serves as a feedstock for agricultural biogas are agroecological sustainable [9, 14, 29]. They prevent environmental degradation such as erosion and soil compaction. They provide lower methane yield [1, 9, 24]. They reduce management effort and cost fertilization. They create a balance in groundwater quality and greenhouse gas emission [5, 9]. This perennial energy crop includes pliscanthus *Miscanthus gigateus*, cup plant *Siliphium perforliatum. L*, wheatgrass *Agropyron elongatum* [14, 37]. Research findings indicate that these perennial energy crops are potentially useful in carbon sequestration [35, 37]. They prevent the use of minimum and conventional tillage which support one or two pass tractors on agricultural land [12, 26]. It retains organic matter in the below and above plants since soil organisms are active and efficient [23, 39]. This practice enhances the formation of complex humid compounds which over time may increase soil organic matter [33, 40]. Some research findings reveal that perennial crops are efficient than annual energy crops such as maize and wheat [6, 11, 19]. Their impact on climate with regards to soil organic matter and greenhouse gas emissions is not beneficial as a perennial energy plant [24, 45]. It was also found that perennial energy crop increases soil organic carbon and nitrogen [29, 37]. Higher microbial biomass and better-developed soil fractal aggregation. it provides a stable organic matter [45]. An experiment was conducted on the distribution and drive of soil organic matter under perennial energy crops. It was found that carbon concentration in the soil cultivated with perennial energy crops was a significantly higher fraction [24, 29]. A significant difference was observed in the perennial crop. Perennial crop such as cup plant and giant-knot weed has higher carbon concentration [24, 29]. It was also found that perennial crop cultivation resulted in higher SOC and elevated bulk density than maize and wheat [24, 29]. It was also recorded that the microbial activities in the soil for the perennial crop were continually [22, 23]. The highest and lowest microbial activities were obtained in the perennial crop than in an annual crop [22, 23]. 1,4 β glucobiosidae shows a positive output with soil organic matter. A consistent increase was observed plot cultivated with perennial energy crops [24, 29]. It was also recorded that the aboveground production depends on the cultivated species. The vegetation period plays a significant role in pre-harvest [22, 45]. This may result in light intensity and internal nutrient distribution for young shoots and roots [24, 29].

5. Digestate a better stimulant for agricultural production

Change in the agro-ecological landscape is a serious challenge. This change was observed in humans, society, and the soil ecosystem [25, 26, 35]. This change brought uncertainty to the soil–plant atmospheric system and variation in the environment. This has resulted in continuous depletion of SOM with ease [12, 13]. Similarly, in an agricultural market survey conducted in Europe. It was discovered that after 1990 a negative phenomenon was recorded in agricultural production [12, 30]. The practice of crop rotation was reduced. Increasing demand for staple food brought a cut down in forage crops by 35% [6, 25]. The production of cereal was increased by 54% and rapeseed by 343% [5, 46]. Likewise, a cut down was documented in animal husbandry by 50% [5, 9, 46]. This was identified as a depleting factor for soil organic matter [5, 9, 46]. The use of mineral fertilizer, organic waste, and manure was found useful but digestate was twice richer compare to other forms of manure [3, 15, 30]. Digestate is made of plants with a large amount of N and P. It is a by-product biogas plant that is capable of providing a high yield of spring crop [3, 15, 30]. It is an excellent fertilizer that can enhance the biological properties of the soil. It is widely used in Europe [2, 17, 30]. Digestate average dry matter content ranges between 1.5% and 46%. Digestate is primarily effective at building up the biological quality of the soil [30]. It can efficiently increase the nutrient quality of the topsoil. It plays a significant role in the carbon pool. SOC content initiates positive or neutral ions with the application of digestate [45, 47]. The amount of organic matter applied to determine the SOC content. The higher the digestate, the larger the SOC vice versa [24, 35]. It was reported that larger particles have a positive influence on SOC content than fine particles [30]. The use of animal droppings as the organic compound was significant [12, 40]. Continuous application of digestate creates a better environment for the decomposition of organic matter. It was also recorded that soil with low quality may prevent the decomposition of organic matter [22, 23, 45]. Accurate dosage of fertilizer and accelerated biological processes in the soil enhance productivity [22, 45, 47]. A decrease in pH and soil sorption complex saturation was found to be an attribute of digestate [30]. This may be due to hydrogen and aluminum ion replacement. A significant reduction was recorded in soil acidification with the application of digestate [3, 15, 30]. The agricultural intensive area was typically high in biological activity. This is significant with soil depth [13, 16, 40]. SOC sequestration is high with the regular application of organic manure. It was recorded that a significant amount of 10 t ha^{-1} of organic manure can increase the SOC stock by 5.5% in 100 years [30, 34, 48]. The use of mineral fertilizer and farmyard manure was also significant to the soil to increasing soil fertility, SOC, and nutrient content [13, 16, 40]. It provides stable nutrients and decreases soil acidity. It was observed that regular application of organic manure provides long-term stable yield and as well improves the quality of the soil [13, 16, 40]. It was found that the use of organic manure and straw is highly valuable to the soil ecosystem [3, 8, 17]. Sufficient application of digestate to the soil can positively influence SOM and SOC content without other manure [15, 30].

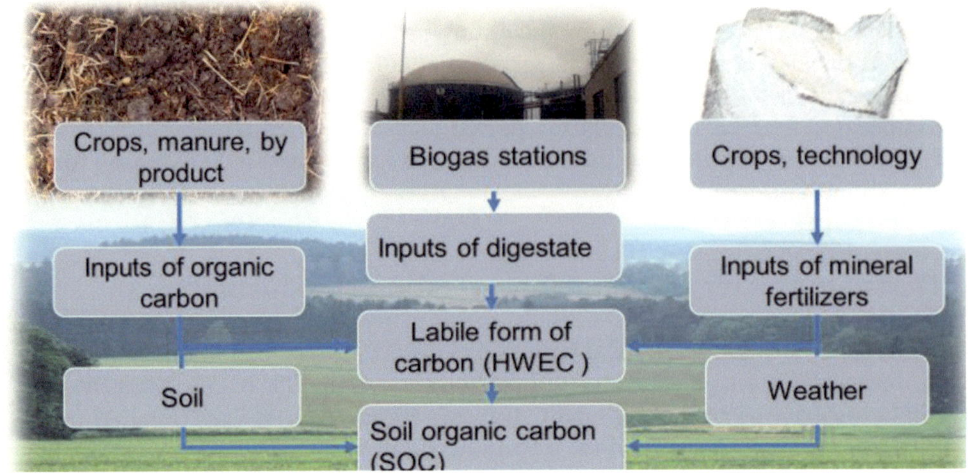

Figure 2.
Digestate production and organic carbon sourced from [49].

Long-term application of digestate maintains SOC content. Postharvest reduces such as straw provide an additional source of organic carbon [4, 24, 30]. A decrease in SOC was recorded in an intensive permanent grassland but SOC and SOM were higher in permanent grassland compare to arable land **Figure 2** [29, 37].

6. The impact of organic carbon on yield

According to research, it was found that the high yield of crops is a result of soil organic carbon. 2% soil organic carbon is critical to the threshold [23, 45]. A value below 2% may have a negative influence on the structural functionality and the asymptotic relationship that exist between soil organic carbon and yield [12, 20]. In other words, the critical threshold for productivity may not be relevant if there is sufficient mineral fertilizer to support crop production [3, 15, 43]. Mineral fertilizer was found to be effective at increasing maize yield [16, 21]. It was reported that mineral fertilizer increases SOC. It improves crop yield and promotes entire growth [2, 11, 31]. It was also found that long-term application of mineral fertilizer significantly increases the SOC content in the 0 – 20 cm soil layer [16, 27, 50]. However, the use of mineral fertilizer is not as effective as the use of straw on farmland [16, 17, 40]. It was discovered in an experiment that the use of straw provides a metabolic substrate for soil microorganisms. Soil porosity and water movement were influenced by straw input [3, 17, 30]. It was also recorded that the use of animal–plant residue increases the SOC content. Straw carbon transformation activates and enhances microbial communities such as bacteria, archaea, protozoa, fungi, and viruses [11–13]. These organisms distribute organic carbon in the complex terrestrial environment [6, 11, 33, 51]. Betaproteobacteria and Gammaproteobacteria were the main genera of *Janthinobacterium*, *Massilia*, *Variovorax*, *Xanthomonas*, and *Pseudomonas* in the early stage of carbon transformation of wheat straw [22, 45, 47]. A positive correlation was recorded relative to SOC to soil microbial structure and diversity. The quality and quantity of SOC are subjected to the metabolic action of the microorganism in the ecosystem [12, 17, 47]. The abundance and structure of microorganisms are generally considered to be essential for the fixation, transport, and accumulation of SOC [17, 22]. They are widely involved in soil processes and functions. Soil microorganisms are effective with the use of organic matter [5, 9].

It supports and shapes the global carbon cycle. It enhances the mechanism that increases SOC and limits the impact of climate change [2, 7, 37]. A limiting factor for straw application is moisture [17, 30]. In cold weather, the accumulation of straw in the soil does not easily decompose. Straw deposition in this case may result in phytotoxic [20, 21, 27]. The use of soil organic carbon is not sufficient enough to influence sustainable intensification to reduce the harm caused by inorganic fertilizer due to eutrophication and greenhouse gas emissions [17, 23, 41]. There is a positive relationship between soil organic carbon and yield starting from the 2% threshold [25, 45, 51]. It provides a reduction in nutrient runoff, drought resistance, and yield stability **Figure 3** [22, 23, 45].

In a finding, the cultivation of maize and wheat uses less than 2% soil organic carbon to area and harvest [2, 12]. It was also found that a continuous cropping system and grazing may result in carbon loss if not properly managed [1, 5, 9]. This practice in other words may improve the yield of maize and wheat due to a large amount of soil organic matter from animal waste [33, 39]. The reduction in nitrogen fertilizer plays a significant role in agricultural land and the ecosystem [16, 20]. It minimizes soil emission of nitrous, eutrophication of water, and efficiency of greenhouse gas [1, 4, 9, 44]. However, as much as soil organic matter is significant to providing nutrients to the soil. This cannot be a direct substitute for mineral fertilizer [16, 20, 40]. The efficiency of soil organic carbon varies between 0.5% and 2.0% to soil properties, climate, and the type of input applied at a point in time [8, 51]. The agricultural input differs in its potential. This efficiency ranges from farmyard manure to sewage sludge to mineral fertilization [13, 20, 27]. It was however concluded that soil organic matter and nutrient provision from agricultural input may be cut down the use of nitrogen fertilizer as input to a large extent [1, 4, 9]. Higher soil organic carbon enhances N input to produce a high yield. Likewise, essential macro and micronutrients are provided through the application of higher levels of soil organic matter [13, 16, 40, 51]. This compensates for soil with limited soil organic carbon concentration [16, 40].

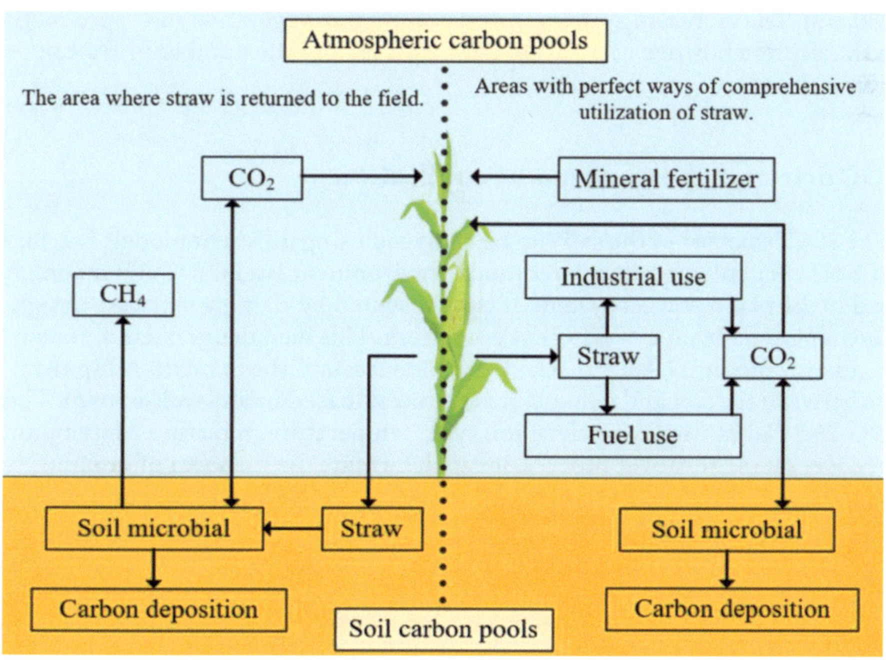

Figure 3.
Carbon pool and straw application sourced from [17].

7. Organic nutrient composition and strawberry yield

The growth and yield of strawberries require macro and micronutrients. Nutrients are provided from the soil as integral support. It is a reservoir for water and nutrient retention. It is a support system for plant growth and root development [36, 52]. Good management practice and strategy for strawberry growth and yield involve the application of compost with synthetic nutrients as a substrate [28, 36, 52]. According to findings, the high growth performance of strawberry cultivars was a result of better nutrient uptake [2, 43]. This nutrient includes Bio plus compost (cocopeat 68.86%, peat moss 11.00%, and zeolite 9.00%, and perlite 11.00% [10]. This is a standardized organic compound used for vegetative growth in some parts of Europe [36]. This nutrient composition is beneficial to microbial development. It supports soil physiochemical properties, the decomposition of organic matter, reduction of eutrophication, and nutrient loss. This is due to organic stipulation and the gradual release of nutrients from this compound into the soil [43]. The nutrient in this organic compound contains a stable proportion of nitrogen and carbon ratio. This however suggests the presence of microbe, humic substances, and high cation exchange capacity [6, 7, 47]. This organic compound plays a significant role in root formation [7, 36, 42]. It is a fiber rich substance. It enhances soil structure [11, 31, 50]. The structural efficiency and stability of this compound lower the impact of global warming and provide the soil with a good carbon sink [31, 33, 51]. In other words, the growth of strawberry cultivar without adequate (NPK) nutrients may have a detrimental effect on vegetative growth [28, 39, 43]. The quality such as sweetness, firmness, and anthocyanin found in strawberries is a result of the optimal application of NPK [28, 39, 43]. The application of this organic compound enhances elongation, carbohydrate, and sugar synthesis [28, 39, 43]. The growth of strawberries using bio plus compost with synthetic nutrients and other growth media was tried in a greenhouse experiment [28, 39, 43]. Results obtained indicated that a significant increase was recorded in the vegetative growth and yield of strawberries compare to other growth media used in the experiment [28, 39, 43]. The increasing growth trend observed in vegetative and reproductive growth reflects a positive correlation with fruit set and the number of fruit per plant [28, 39, 43].

8. SOC determination and model application

The SOC behavior in the soil can be analyzed using different models [17, 18, 49]. The CENTURY soil organic matter model environment is a FORTRAN model. It is used in the plant–soil ecosystem. It is represented by C. It is a software design for nutrient dynamism in a different ecosystem. This includes grassland, forest, crop, and savannah [17, 18, 49]. The EPIC model establishes a relationship that exists between the soil and climate. It is a process-based model well-known. The Roth C-26.3 can be used to analyze soil type, temperature, moisture content, and plant cover on the turnover process. It can determine the turnover of organic carbon in non-waterlogged soil [17, 18, 49]. Some of the required parameters for analyzing SOM and SOC were weather, sowing procedures, plan of work on the different fields, fertilization etc. This information is based on different algorithms [17, 18, 49]. STIC model was also noted for a good approximation of water requirement. However, the use of multiple linear regression (MLR) was noted for its multidimensional functions [17, 18, 49]. MLR is capable of modeling soil properties after the application of organic manure. MLR can be used accomplished to the relationship between SOC stock and other soil properties in any region of the world.

It can provide a relationship between liable C form, soil properties, and management practices. The use of MLR was found efficient and reliable [17, 18, 49].

9. Conclusions

Global warming is an atmospheric challenge that limit the efficiency of food production across the globe. It determines carbon sink its availability in the soil. Agricultural farmland is an important pool for carbon sink and deposit. Organic matter enhances water retention and nutrient built up in the soil. It cut down the input mineral fertilizer. It provides carbon stability and crop yield. Soil carbon accumulation largely depend on the rate at which biomass decomposes. Approximately 40% - 60% carbon are lost through conventional tillage. It regulates the extent to which chemical constituent is released into the soil. Plowing exposes the soil to direct insolation which may support rapid deposition. Erosion prevents direct oxidation of carbon into CO2. Carbon sequestration is high with crop that have abundant residue. The no-till practice perform the function of deepening carbon into the soil. A land cover with forest is significant to accumulating carbon and nitrogen in the soil. The nutritional quality of crop is lowered with increased atmospheric CO2. The growth of the plant and natural storage of carbon in the soil serves as defense against climate change. Peatland is capable of striking balance between vegetation and carbon reservoir. Perennial crop prevents environmental degradation such as erosion and soil compaction. Digestate is an excellent fertilizer that can enhance the biological properties of the soil. The higher digestate, the larger the SOC vice versa. Straw carbon transformation can activate microbial community such as bacteria, archaea, protozoa, fungus, and viruses. Multiple linear regression is a multidimensional software. It built a relationship between SOC stock and soil properties. Other software used for determining SOC are CENTURY soil organic matter model, EPIC model, Roth C-26.3 and the STIC model.

Author details

Alabi Olusoji David
Mangosuthu University of Technology, Durban, South Africa

*Address all correspondence to: oludavidsj@gmail.com

References

[1] Diffenbaugh, N. S. Burke, M. (2019) Global warming has increased global economic inequality. Proc. National. Academic Science. U.S.A. 116, 9808-9813.

[2] Jiang, C.M. Yu, W.T., Ma, Q. Xu, Y.G. Zou, H. (2017): Alleviating global warming potential by soil carbon sequestration: A multi-level straw incorporation experiment from a maize cropping system in Northeast China. Soil and Tillage Research, 170: 77-84.

[3] UNCCD, Land degradation, poverty and inequality (2019). https://www.unccd.int/ publications/land-degradation-poverty-and-inequality. Accessed 11 August 2020.

[4] Duan, P. Fan, C. Zhang, Q. Xiong, Z. (2019): Overdose fertilization induced ammonia-oxidizing archaea producing nitrous oxide in intensive vegetable fields. Science of the Total Environment, 650: 1787-1794.

[5] De Paul Obade V. (2019): Integrating management information with soil quality dynamics to monitor agricultural productivity. Science of the Total Environment, 651: 2036-2043.

[6] Hatfield, J. L. Sauer, T. J. & Cruse, R. M.: (2017). The Forgotten Piece of the Water, Food, Energy Nexus, Adv. Agron., 143, 1-46,

[7] Jia, Z. Kuzyakov, Y. Myrold, D. Tiedje, J. (2017): Soil organic carbon in a changing world. Pedosphere, 27: 789-791.

[8] Rao, C.S. Indoria, A.K. Sharma, K.L. (2017) Effective Management Practices for Improving Soil Organic Matter for Increasing Crop Productivity in Rainfed Agroecology of India. Current Science, 112, 1497-1504.

[9] Drake, T.W. Podgorski, D.C. Dinga, B. Chanton, J. Six, J. Spencer, R.G.M. (2019): Land-use controls on carbon biogeochemistry in lowland streams of the Congo Basin. Global Change Biology, 26: 1374-1389.

[10] Wei, H. Liu, C. Ryong Jeong, B. (2020). An optimal combination of the propagation medium and fogging duration enhances the survival, rooting, and early growth of strawberry daughter plants. Agronomy 10 (4), 557.

[11] Hijbeek, R. van, Ittersum, M. K. ten Berge, H. F. M. Gort, G. Spiegel, H. & Whitmore, A. P.:(2017). Do organic inputs matter – a meta-analysis of additional yield effects for arable crops in Europe, Plant Soil, 411, 293-303.

[12] Wang, , J. Zhang, H. Dong, Y. Zhang, Y. (2019) The role of iron oxides in the preservation of soil organic matter under long-term fertilization. Journal Soils Sediments, 19, 588-598.

[13] Zhang, X. Fang, Q. Zhang, T. Ma, W. Velthof, G.L. Hou, Y. Oenema, O. Zhang, F. (2020). Benefits and trade-offs of replacing synthetic fertilizers by animal manures in crop production in China: A meta-analysis. Global Change Biology. 2020, 26, 888-900.

[14] FAOAQUASTAT, AQUASTAT. Food Agriculture Organization, (United Nations, 2019), http://www.fao.org/nr/water/aquastat/data/query/index.html?lang = en. Accessed 11 August 2020.

[15] 15.Shi, R.-Y. Liu, Z. D. Li, Y. Jiang, T. Xu, M. Li, J.Y. Xu, R.-K. (2019). Mechanisms for increasing soil resistance to acidification by long-term manure application. Soil Tillage Res., 185, 77-84.

[16] Zhang, X. Fang, Q. Zhang, T. Ma, W. Velthof, G.L. Hou, Y. Oenema, O. Zhang, F. (2020). Benefits and trade-offs of replacing synthetic fertilizers

with animal manures in crop production in China: A meta-analysis. Global Change Biology, 26, 888-900.

[17] Qiuju Wang, Xin Liu, Jingyang Li, Xiaoyu Yang, Zhenhua Guo (2021). Straw application and soil organic carbon change. A meta-analysis. *Soil and Water Research, 16, 2021 (2): 112-120.*

[18] Šimon, T. Madaras, M. (2020). Chemical and Spectroscopic Parameters Are Equally Sensitive in Describing Soil Organic Matter Changes after Decades of Different Fertilization. Agriculture, 10, 422.

[19] Chabbi, A., Lehmann, J. Ciais, P. Loescher, H. W. Cotrufo, M. F. Don, A., SanClements, M. Schipper, L., Six, J. Smith, P., and Rumpel, C.(2017). Aligning agriculture and climate policy, Nature Clim. Change, 7, 307-309,.

[20] Bhardwaj, A. K., Jasrotia, P., Hamilton, S. K., and Robertson, G. P.:(2011) Ecological management of intensively cropped agro-ecosystems improves soil quality with sustained productivity, Agricultural Ecosystem Environment., 140, 419-429.

[21] Bauer, A. and Black, A. L. (1992). Organic carbon effects on available water capacity of three soil textural groups, Soil Science. Society. America. Journal, 56, 248-254.

[22] Paul E. (2016): The nature and dynamics of soil organic matter: Plant inputs, microbial transformations, and organic matter stabilization. Soil Biology and Biochemistry, 98: 109-126.

[23] Poulton, P. Johnston, J. Macdonald, A. White, R. & Powlson, D. (2018). Major limitations to achieving "4 per 1000" increases in soil organic carbon stock in temperate regions: Evidence from long-term experiments at Rothamsted Research, United Kingdom, Global Change Biology., 24, 2563-2584,

[24] Chimento, C. Almagro, M. & Amaducci, S. (2016). Carbon sequestration potential in perennial bioenergy crops: The importance of organic matter inputs and its physical protection. Global Change Biology-Bioenergy, *8*, 111-121.

[25] Bentsen, N.S. Larsen, S Stupak, I. (2019). Sustainability governance of the Danish bioeconomy—The case of bioenergy and biomaterials from agriculture. Energy Sustainability Society, 9, 40.

[26] Brain, R.A. Anderson, J.C. (2019). The agro-enabled urban revolution, pesticides, politics, and popular culture: a case study of land use, birds, and insecticides in the USA. Environmental Science and Pollution Research International, 26: 21717-21735.

[27] Rasmussen, C. Heckman, K. Wieder, W. R. Keiluweit, M. Lawrence, C. R. Berhe, A. A. Blankinship, J. C. Crow, S. E. Druhan, J. L. Pries, C. E. H. Marin-Spiotta, E. Plante, A. F., Schädel, C. Schimel, J. P. Sierra, C. A., Thompson, A. & Wagai, R. (2018). Beyond clay: towards an improved set of variables for predicting soil organic matter content, Biogeochemistry, 137, 297-306.

[28] Elanchezhian, A. Khan, F. Basak, J.K. Park, J. Okyere, F.G. Lee, Y.J. Kim, H.T. (2019). Analysis of water retention capacities of various compost and its relationship to the strawberry moisture level. In: International Symposium on Advanced Technologies and Management for Innovative Greenhouses, 1296, pp. 899-906.

[29] Ferchaud, F. Vitte, G. & Mary, B. (2016). Changes in soil carbon stocks under perennial and annual bioenergy crops. *Global Change Biology-Bioenergy*, *8*, 290-306.

[30] Šimon, T. Kunzová, E. & Friedlová, M. (2015). The effect of digestate, cattle

slurry and mineral fertilization on the winter wheat yield and soil quality parameters, Plant Soil Environment, 62, 522-527.

[31] Hasnain, M. Chen, J. Ahmed, N. Memon, S. Wang, L. Wang, Y. Wang, P. (2020). The effects of fertilizer type and application time on soil properties, plant traits, yield, and quality of tomato. Sustainability 12 (21), 9065.

[32] Hannam, K.D. Midwood, A.J. Neilsen, D. Forge, T.A. Jones, M.D. (2019): Bicarbonates dissolved in irrigation water contribute to soil CO2 efflux. Geoderma, 337: 1097-1104.

[33] Hu, J. Guo, H. Wang, X. Gao, M.T. Yao, G. Tsang, Y.F. Li, J. Yan, J. Zhang, S. (2019): Utilization of the saccharification residue of rice straw in the preparation of biochar is a novel strategy for reducing CO_2 emissions. Science of the Total Environment, 650: 1141-1148.

[34] Missirian, A. Schlenker, W. (2017) Asylum applications respond to temperature fluctuations. Science 358, 1610-1614.

[35] Bi, Y. Cai, S. Wang, Y. Xia, Y. Zhao, X. Wang, S. Xing, G. (2019). Assessing the viability of soil successive straw biochar amendment based on a five-year column trial with six different soils: Views from crop production, carbon sequestration, and net ecosystem economic benefits. Journal of Environmental Management, 245: 173-186.

[36] Khan, F. Okyere, F.G. Basak, J.K. Qasim, W. Park, J. Arulmozhi, E. Kim, H.T. (2019). Comparison of different compost materials for growing strawberry plants. In: International Symposium on Advanced Technologies and Management for Innovative Greenhouses: GreenSys 2019, 1296, pp. 869-876.

[37] Gansberger, M. Montgomery, L.F.R. & Liebhard, P. (2015). Botanical characteristics, crop management, and potential of *Silphium perfoliatum* L. as a renewable resource for biogas production: A review. *Industrial Crops and Products*, *63*, 362-372.

[38] Kroon, F. J. Thorburn, P. Schaffelke, B. Whitten, S. (2016). Towards protecting the Great Barrier Reef from land-based pollution. Global. Change. Biology. 22, 1985-2002.

[39] Madhave, B.G.K. Khan, F. Bhujel, A. Jaihuni, M. Kim, N.E. Moon, B.E. (2021). Influence of different growth media on the growth and development of strawberry plants. Cell press, Heliyon 7 e7071.

[40] Wang, *P. Wang*, J. Zhang, H. Dong, Y. Zhang, Y. (2019). The role of iron oxides in the preservation of soil organic matter under long-term fertilization. Journal Soils Sediments, 19, 588-598.

[41] Ritchie, H. Roser, M. Clean water. Our World in Data (2020). https:// ourworldindata. org/water-access. Accessed 11 August 2020.

[42] Bauer, A. & Black, A. L.: Quantification of the Effect of Soil Organic Matter Content on Soil Productivity, Soil Science Society America Journal, 58, 185, 1994.

[43] Vandecasteele, B. Debode, J. Willekens, K. Van Delm, T. (2018). Recycling of P and K in circular horticulture through compost application in sustainable growing media for fertigated strawberry cultivation. European. Journal Agronomy. 96, 131-145.

[44] Skrypchuk, P. Zhukovskyy, V. Shpak, H. Zhukovska, N. Krupko, H. (2020). Applied Aspects of Humus Balance Modeling in the Rivne region of Ukraine. Journal Ecology Engineering, 21, 42-52.

[45] Oldfield, E.E. Bradford, M.A. Wood, S.A. (2019). A global meta-analysis of the relationship between

soil organic matter and crop yields. Soil, 5, 15-32.

[46] De La Cruz, M.L. Conrado, I. Nault, A. Perez, A. Dominguez, L. Alvarez, J. (2017): Vaccination as a control strategy against Salmonella infection in pigs: A systematic review and meta-analysis of the literature. Research in Veterinary Science, 114: 86-94.

[47] Menšík, L. Hlisnikovský, L. Pospíšilová, L. Kunzová, E. (2018). The effect of application of organic manures and mineral fertilizers on the state of soil organic matter and nutrients in the long-term field experiment. Journal Soils Sediments, 18, 2813-2822.

[48] Maillard, É. Angers, D.A. (2014). Animal manure application and soil organic carbon stocks: A meta-analysis. Global. Change Biology, 20, 666-679.

[49] Václav Voltr, Ladislav Menšík, Lukáš Hlisnikovský, Martin Hruška, Eduard Pokorný and Lubica Pospíšilová. (2021). The soil organic matter in connection with soil properties and soil inputs. Agronomy, Mdpi, 11, 779.

[50] Ramankutty, N. Hertel, T. Lee, H. L. & Rose, S. K. (2007). Global Agricultural Land Use Data for Integrated Assessment Modeling, in Human-induced climate change: An interdisciplinary assessment, Cambridge University Press, New York.

[51] Hickman, J.E. Huang, Y. Wu, S. Diru, W. Groffman, P.M. Tully, K.L. Palm, C.A. (2017): Nonlinear response of nitric oxide fluxes to fertilizer inputs and the impacts of agricultural intensification on tropospheric ozone pollution in Kenya. Global Change Biology, 23: 3193-3204.

[52] Raja, W.H. Kumawat, K. Sharma, O. Sharma, A. Mir, J. Nabi, S.U. Qureshi, I. (2018). Effect of different substrates on growth and quality of Strawberry cv. chandler in soilless culture. Pharma Innov. J 7, 449-453.

Chapter 3

Organic Farming for Sustainable Agriculture Using Water and Soil Nutrients

Sanjeevakumar M. Hatture, Pallavi V. Yankati, Rashmi Saini and Rashmi P. Karchi

Abstract

The agricultural community/farmers are struggling to obtain higher rate of yield due to lack of poor knowledge about the soil and water nutrients and suitability of the organic crop for the soil. Most of the farmers use excessive chemical fertilizers in-order to increase productivity of their yield, without aware of side effects. The excess usage of chemical fertilizers by the farmers will have impact on the quality, fertility, and salinity of the soil. To overcome these issues and to promote Digital Agriculture concept we propose an IoT enabled sensor system for monitoring soil nutrient [NPK] and pH of irrigation water to reduce the manual laboratory method of testing and get the results via mobile application and to promote organic farming in the agricultural field. Smart organic farming based mobile application will further process these nutrients value to predict and suggests the suitable crop to grow and the usage of appropriate amount of fertilizer to maintain the soil fertility there by achieving optimum usage of chemical fertilizer because continuous and wrong usage of these chemical fertilizer have a harmful effect not only on soil but also on crops, we consume leading to unhealthy human life. The proposed mobile application also helps in establishing the connection between farmers and Agricultural Produce Market Committee (APMC) in order to avoid fragmentation of profit shares and attain Pricing uncertainty and marketing of the yields by avoiding the middle man. APMC is a state government body which ensures safeguard to the farmers from exploitation by large retailers and suggest the kind of crop to be grown with organic farming. India is well known to produce organic fertilizer which is produced by the waste of slaughterhouses, plant and animal residues, biological products and other natural resources. Thus, the proposed work helps the farmers in adopting stress-free organic farming practice by self-testing their field soil parameters for generating quick soil analysis reports and also helps in connecting with APMC to know the suitable crop for their agriculture land based on the soil and water analysis (SWA) report, dispensing the required amount of organic fertilizer to the soil based on soil and water nutrients analysis using IoT enabled sensor, funding/insurance to the crops in case of occurrence of unpredictable natural disaster in future and direct marketing facility without middle man and maintain sustainable agriculture. In the present era, the industry is at 5.0 levels but agricultural production is still at 2.0 levels. In this chapter a methodology for sustainable agriculture and increase the organic yield of the organic farming using the mobile and IoT technological approaches is presented. A former can obtain the advice and other information for growing the organic crop, organic certification,

 IntechOpen

pricing for the organic yield, selling and other activities by using mobile application in his/her local language. By the proposed work with the ease of mobile application the farmers can perform self-test of their field parameters for generating quick soil and water analysis report, predicts and suggest the suitable organic crop, obtaining the suitable pricing by the APMC and organic certification and agreement to meet the sustainable agriculture. Further the soil fertility of the organic farm can be monitored using IoT enabled sensors which are remotely connected with the mobile application. The experimentation is performed at different agriculture fields with organic farming at six geographical separated villages at Bagalkot district of Karnataka state, India. The different agricultural lands with variety of soil samples is tested to measure the soil parameter such as moisture, temperature, humidity and NPK nutrient values. The pH value of the irrigation water is also determined including borewell, pond, rain, river water etc. available in the reservoirs and promising sustainability in the organic yield is obtained.

Keywords: Organic farming, Smart agriculture, Soil and Water Nutrients, IoT, Sensor, APMC

1. Introduction

In the era technology enable modern agriculture, the people are keen about the organic farming due to environmental friendliness and more profitable agriculture. The people are diverting towards the organic farming as it uses less chemical-intensive pesticides. At present, agriculture besides farming includes forestry, fruit cultivation, dairy, poultry, mushroom, bee keeping etc. Today marketing, processing and distribution of agricultural products etc. are all accepted as a part of modern agriculture. Thus, agriculture can be defined as the art or science of production of crops and livestock on farm. There are three major types of agricultural practices in India, among them first one is Subsistence Agriculture which means farmers will grow crops for their self needs. Second is Commercial Farming, here farmers will grow crops for earning money by selling the grown crops and the last one is Sustainable Agriculture which is the efficient production of safe, high-quality agricultural product, in a way that protects and improves the natural environment, the social and economic conditions of the farmers.

Sustainable agriculture is farming in sustainable ways meeting sustainability criteria with adaptation of technological approaches. Sustainability criteria can be generally defined as the requirement to the sustainable quality of a product and its sustainable production, which have to be fulfilled in order to acquire a sustainability status. It mainly depends upon three dimensions those are Social Sustainability, Economic Sustainability and Environment Sustainability. The **Table 1** depicts the sustainability criteria for attaining the higher yield of the organic crop and sustainable agriculture based on three dimensions.

Earlier days there was no scarcity for food and water as it was used in efficient manner. But now a day there is scarcity of food and water as usage of technology and medicines have been increased due to increase in population growth of the country. To satisfy their needs there is a need in increase in crop production, to produce more crops more amount of fertilizer is being used as it develops more crops in short period of time. Eating these crops grown using more chemical fertilizer will be harmful to the human beings as it causes savior diseases. Hence it is must and should necessary for farmers to know about the usage of fertilizer for cropping and this fertilizer usage depends upon the amount of nutrient present in

Sustainability dimension	Criteria
Social sustainability	• Creating self-reliance in organic production • Empowering rural communities through partnerships with other farmers to form groups (providing participatory guarantee system and the strengthening of social organization) • Increasing farm employment • Improving employment opportunities, particularly in rural areas
Economic sustainability	• Improving efficiency in areas with low inputs (pesticides, herbicides, etc.) • Reducing crop damage and Reducing risk of production • Satisfying farmers from an economic perspective • Added value of organic products through marketing activities and processing • Efficient usage of resources more efficiently (to minimize the use of non-renewable resources) • Affordable than traditional agriculture (due to lower variable costs of inputs, identical Fixed costs and higher prices of organic products) • Greater economic profitability due to the use of domestic inputs • Enhancing the overall performance of the farm in unit area
Environmental sustainability	• Maintaining soil fertility in the long-term • Improving soil structure • Increasing soil water retention potential • Producing food without chemicals • Using environmentally friendly production methods (such as the use of animal products, intercropping, use of mulch, using natural pesticides, crop residue returned, green manure, compost, crop rotation) • Reducing environmental pollution (pollution reduction of water, soil, and air)

Table 1.
Important sustainability criteria in organic farming.

the soil. The other issue is that farmers do not test their soil and hence they do not know the level of nutrients in soil. Nutrients are more essential for crop's growth as well as for human health. The deficiency of nutrients content in the crop will cause disease to plant such as stunted growth, leaf yellowing or browning and chlorosis. The human beings eating these nutrient deficient crops are also exposed to severe diseases such as growth disorders, increased susceptibility for infections, skin rashes and immune suppression. These nutrients are classified into micro and macro nutrients. Totally soil consists sixteen nutrients those are carbon, hydrogen, oxygen (which are supplied by air and water), calcium, magnesium, Sulfur (which are secondary nutrients), nitrogen, phosphorous, potassium (which are Macronutrients), boron, chlorine, copper, iron, manganese, molybdenum, and zinc (which are Micronutrients) among them the three major nutrients are Nitrogen (N), Phosphorus (P), Potassium (K).

In organic farming, a proper soil testing will help the farmers to get the correct amount of nutrients present in the soil and to choose the right crop for growing. For achieving the sustainable agriculture, proper management of essential soil nutrients play a vital role. Technology such as IoT and Machine Learning plays an expedient role for improvement of environment and for achieving the economic goals.

Farming is one of the most important occupation since the beginning of civilization. Even though farmer is rich enough to serve the food for every individual of the society but poor himself due to improper pricing for their cultivated crop due to the intervention of the third party in marketing the agriculture products. Since, the farmer's family is depending on cultivation, if crops are not as per the demands, then farmer's life becomes miserable hence farming occupation is low admired job. On the other hand, besides of having grown good quality crops or agriculture products they are enable to sell their crops due to occurrence of unexpected global health crisis such as Covid-19 (Novel Coronavirus). Because of the disruptions caused by the Covid-19 outbreak the government has imposed a countrywide lockdown to stop the spread of the coronavirus pandemic. Wherein all internal and external transport has been banned strictly this impacts the farmers with heavy losses due to lack of medium to transport. In such situation adopting to the smart farming techniques is beneficiary, in which everything including buying and selling of agriculture products are done through online sitting in home itself so that it will be profitable to both.

Focusing on encouraging innovation in agriculture, smart farming is the best answer to the problems stated above. This Smart Farming is a concept of agricultural management using modern Information and Communication Technologies such as IoT and Machine Learning to increase the quantity and quality of products. The internet of things (IoT), is a system of interrelated computing devices which are provided with unique identifiers (UIDs) and the ability to transfer data over a network without requiring human-to-human interaction. An IoT ecosystem consists of web-enabled smart devices such as sensors to measure the nutrients present in the soil. IoT devices share the sensor data they collect by connecting to an IoT gateway or other edge device where data is either sent to the cloud to be analyzed or analyzed locally. IoT can also make use of other technology such as machine learning to aid in making prediction based on collected data. IoT can benefit farmers in agriculture by making their job easier.

Machine Learning is the most popular technique of predicting the future or classifying information to help people in making necessary decisions. These algorithms learn from the past instances of data through statistical analysis and pattern matching. Then, based on the learned data, it provides us with the predicted results.

1.1 Overview of proposed work

The organic farming products require proper marketing facilities to the farmers. Information and Communication Technology play a vital role in connecting people all over the world in a fraction of second and has proven impact on their everyday life. It greatly effects in improving the lives of farmers in India. The smart phone is the most commonly used ICT tool all over world, and farmers have started to use mobile phone for communication in their native languages. So, it can help them in direct communication involving the farmers and buyer or customers. By investigating and considering the entire problem faced by farmers the proposed work has introduced smart farming based android mobile application which tests the soil nutrients using the IoT enabled light and color sensors. Farmers need to install the application on their mobile phones so that they can easily get/view the real time soil nutrient [N, P, K] values on their mobile phones. To determine the soil nutrients, value the hardware set up is made which includes sensor, Ardino microcontroller, serial port, cable connector, regulated power supply, soil sample.

The mobile application also helps in establishing the connection between farmers and Agricultural Produce Market Committee (APMC) in order to attain Pricing uncertainty and marketing of the organic farming yields by avoiding the middle man. APMC is a state government body which ensures safeguard to the farmers from exploitation by large retailers. APMC admin can view the details of the registered farmers like land area, land number etc. The registered organic farmers of the mobile application get the facility to share the soil and water nutrients reports to the APMC authorities. Further based on the reports, the APMC authorities can recommend the suitable crop to be grown to the farmers and issue the organic certification. And also offers an agreement proposal to the farmer which consists of purchasing price and insurance for their yields thus promoting direct market and distribution facility to the farmers. The organic farming product are costly as it requires more workers and is produced in smaller amounts. According to this agreement, if in situations like Droughts, flood and any other Natural Disasters occurs the APMC authority should pay a minimum amount to the farmers which is fixed by the government. The mobile application also provides the facility for the farmers to get the awareness and usage of new tools and technology, and suggestive measure for the organic farming.

1.2 Scope and objective

The objective of the proposed mobile application used for the organic farming is to,

- Assist the farmers to communicate with APMC for buying and selling of their organic products through online by establishing a link between them.
- Support regional language, hence it increases the usability of mobile application among the farmers.
- Automate the processes of testing the agriculture land to determine the pH of irrigation water and soil nutrients (NPK).
- Facilitate the farmers to share the soil test report with the APMC.
- Suggests the suitable crop based on the uploaded soil test report.
- Facilitate the organic farming with an organic certification, fair pricing for their organic yields without intervention of any middle man through smart farming practices.

1.3 Motivation

The day by day increase in population growth, there is huge need of crop production this crop production mainly depends on the nutrients present in the soil, pH of irrigation water supplied and location specific crop cultivation. Hence it is required for farmers to select suitable crop by monitoring water pH and soil nutrients of the agricultural land. The main motivation of this project is to automate the process of soil nutrient and pH of irrigation water testing and develop an advisory mobile application for farmers to suggests suitable crop based on analysis made on the soil test report to increase crop yield. Farmers usually starts work early during planting and work till harvesting of crops because of their hard work everyone has their daily bread on table. His/her life can be made little better by establishing communication link between APMC authority and farmer to ensure fair marketing price by providing direct selling and buying option through the developed mobile application. Thus, achieving minimum profit out of cultivation. So, farmer's life can be made a tension free by guarantying buying of the agriculture product before they start cultivation through this mobile app and provide all the necessary information for farming or cultivation hence, this mobile application is developed.

1.4 Organization of chapter

The entire project chapter is organized in six sections. **Section 1** Introduces the system, and presents overview, scope and objectives and motivation of the proposed work. **Section 2** provides the summary of the existing work and survey of various literatures. **Section 3** describes the proposed model with module description, workflow and advantages of system. **Section 4** discusses the system design and description of each module of the system. Finally, **Section 5** depicts the results and discussion, comparison and method of increasing the accuracy with Graphical result analysis.

1.4.1 Section 2: Literature survey

Organic farming started along with the human civilization but it was totally dependent on nature. Due to increase in the population, there is higher demand for the agricultural products also increased. In-order to cope with this demand farmers started using fertilizers. Indiscriminate use of pesticides and chemicals polluted soil, water and whole ecosystem. Off late our farmers from different parts of India are switching over to sustainable agriculture practice in fields by adopting smart farming techniques as part of Digital India campaign to promote Digital Agriculture.

Exhaustive literature survey is carried out to determine the soil primary nutrients and pH value of the irrigation water of the agriculture field, the technologies employed for the agricultural activities, the mobile applications etc. The research work is enlisted in the **Table 2**.

Location specific cultivation of crop provides better yield and selecting the best suitable crop based on the soil parameter using ICT enabled Machine Learning technology results in high accuracy and sustainable agriculture. Some of recent works done in this direction are summarized and enlisted in **Table 3**.

Reference	Proposed work	Sensor used
[1]	IoT sensor for monitoring the soil moisture content with automatic irrigation to save excess water usage.	Soil Moisture Sensors
[2]	Develop an intelligent System to monitor the crop by using drones fitted with a camera eye to record images of crops in a scheduled time	Computer Vision System to monitor the crop
[3]	Monitoring of the field and to provide the proper fertilizers depending upon the soil nutrients.	Soil moisture, pH, PIR, ultrasonic and color sensor
[4]	The soil characters are determined and monitored using various sensors. The data determined are saved in server for later use.	pH sensor, soil moisture sensor, infrared sensor
[5]	Aim is to ease the farmers work by providing automatic planting of crops and monitor the field remotely using ICT enabled tools and technology	Sensor to capture the crop image, and PC to monitor it.
[6]	Soil is foundation for any plant growth. The soil components are measured using various sensors. The real time data are put away and refreshed in cloud worker for further processing	pH sensor, soil moisture sensor, infrared sensor, Humidity sensor
[7]	The proposed of equipment module is constructed utilizing the raspberry-pi as a minicomputer to process all live info information from the sensor to give fundamental data about land information as yield.	Raspberry Pi USB GPS programming.

Reference	Proposed work	Sensor used
[8]	Mainly focuses on the strategy to protect the crops during inevitable condition and inculcating technology implementation to promote smart Agro-environment.	GSM and DTMF technology
[9]	Provides systematic agriculture monitoring irrigation techniques which consists of sensor to sense water level of soil and different other parameters of soil.	Soil Moisture Sensor, temperature and humidity sensor.
[10]	Crop productivity mainly depends upon soil and water parameters, the proposed system helps in monitoring the soil and water parameter using several sensors.	Soil Sensor and Infrared sensors
[11]	Objective is to improve the crop production as well as increase the GDP of nation using IoT ecosystem to bring new beginning in the agriculture filed.	Soil moisture, DHT11 sensor and Light sensor
[12]	Goal is to collect live data of agriculture and environment to provide necessary advice on weather condition, crops selection etc. using messages through Short Massaging Service (SMS)	Sensing local agricultural parameters using several soil sensors
[13]	Presents complete automatic drip irrigation system for the agriculture fields by determining the pH, moisture content and the nutrient content of the soil	ARM9 processor, GSM module
[14]	Proposed model aims at restoring the levels of nutrients by monitoring the NPK present in the soil using sensor. And developed a system to provide appropriate amount of fertilizers.	soil fertility sensor, pH sensor and color sensor
[15]	Aims at developing portable handheld device for estimating the nutrient content of the soil and a mobile app for further analysis and comparison.	EC sensor, pH sensor and color sensor

Table 2.
Summary of the literature review.

1.4.2 Summary of the literature survey

In the literature survey, several techniques are proposed by the authors by covering various areas of the agriculture including usage of different technology such as Internet of Things (IoT) technique for determining and monitoring soil nutrients, Drones for automation of crop management and identifying harvesting period of crop. Further Machine Learning calculations, for example, k-Nearest Neighbor (k-NN), Support Vector Machines (SVM) and Artificial Neural Network (ANN) have been utilized for foreseeing the harvest yield. The mobile applications for the educated farmers in monitoring the agriculture activities including suggestions, notifications, APMC marketing, direct marketing etc.

1.5 Problems in the existing system

No proper guidance about the soil nutrients is available to the farmer, organic crop/food requirement in the market, non-support of local language in the application, lack of awareness about the location specific crop cultivation and less profit due to occurrence of third party in marketing of agricultural products [26].

1.6 Issues and challenges

Some of the issue and challenges of the existing techniques proposed by various authors and reported in the literature are identified and are enlisted in the following:

- **Lack of Knowledge About the Soil and Water Parameters:** Soil and water are the most important factors for any crops to grow having less knowledge about the soil nutrients and pH of irrigation water will reduce the agriculture profit or in sometimes it results in the crop failure.

- **Lack of Concern Towards New Technology:** The transformation to new innovation is still amazingly poor and the greater part of the farmers are yet uninformed with respect to such progressions.

- **Negligence Related Natural Resources:** inattention towards natural resources like water, soil etc. will lead to changes in the agriculture productivity and natural assets.

- **Over-Dependence on Traditional Crops than Organic Crops:** Earlier farmers follow the traditional cropping system but nowadays due to increased use of fertilizer and same cropping system will reduce the nutrients level of the soil. It results in the reduce crop production and produce low quality crops. Hence selection of crops should be based on the soil parameter and type of soil.

Reference	Proposed work	Algorithm used
[16]	Selection of location specific crop plays an important parameter in getting high and stable yield.	Information Fuzzy Network and Data Mining Techniques
[17]	Objective is to predict and suggest the end user/farmer about the crop yield with proper recommendation of fertilizer ratio.	Backpropagation algorithm
[18]	Developed mobile application to support the farmers in getting job notification, search investors across the country and also identify the ripeness of fruit for banana and grapes	Support Vector Machine Classifier
[19]	To increase GDP of nation, a mobile application is developed to predict suitable crop and its future price based on the user input location.	Long Short-Term Memory (LSTM) recurrent neural networks (RNNs)
[20]	developed a mobile application to predict future weather condition of remote area.	Random Forest Classification
[21]	Goal is to build a hardware device to measure soil fertility and accordingly a mobile app is developed to predict and suggest crop and fertilizer plan to farmer.	Support Vector Machine Classifier
[22]	Developed model to classify the soil based on the land type and accordingly predict and suggest the suitable crop to farmers	k-Nearest Neighbor (k-NN), Bagged Trees, Gaussian kernel based Support Vector Machines (SVM).
[23]	Objective of the proposed mobile app is to provide chat forum for farmers to share and get suggestion from experts about crop and fertilizer recommendation.	Predictive Analytics
[24]	Objective is to predict the most profitable crop using data analytics techniques	Multiple Linear Regression
[25]	Predicts the crop yield from the available historical data of Tamil Nadu weather condition.	Random Forest

Table 3.
Location specific cultivation of crop.

- **Poor Marketing:** Absence of market offices and helpless government guidelines, and so forth., make it practically incomprehensible for farmers particularly little scope farmers to showcase their homestead produce. Improved market offices and great government guidelines can go far in helping helpless farmers market and benefit from their harvests.

Hence in the proposed work design and develop of organic farming based advisory mobile application to test soil nutrients and pH of irrigation water of agricultural lands there by providing suggestive measures to the farmers for the type of organic crops to be grown, usage of the fertilizers in-order to increase the soil fertility and yield of the land.

2. Proposed system

The smart farming for sustainable agriculture using soil and water nutrients is a IoT enabled system exploring the android mobile application. The proposed work tries to overcome the problems of the existing system and to lower the farming cost a smart farming based android mobile application is proposed to help the farmers in monitoring and maintenance of agricultural land to get high yield, crop selection, promoting organic farming to achieve sustainability in agricultural activities is depicted in **Figure 1**.

2.1 Block diagram

The proposed organic farming based mobile application provide separate login facility is to both Agricultural Produce Market Committee (APMC) admin and farmer. It also helps farmer to communicate with Agricultural Produce Market Committee (APMC) to avail the facilities provided by the APMC such as uploading

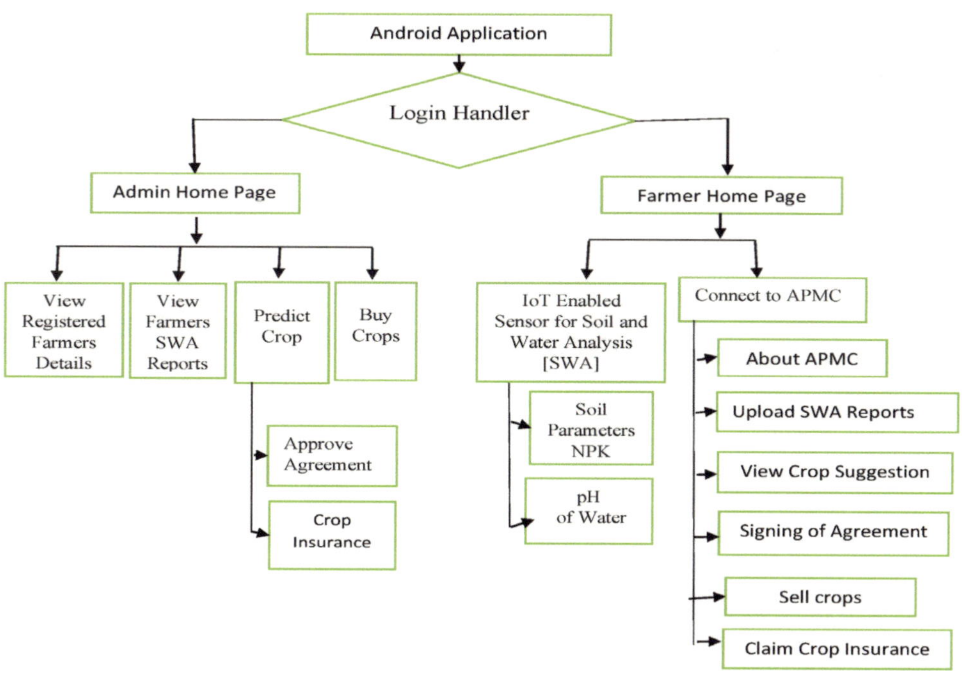

Figure 1.
Flow of the processes in proposed model.

soil and water analysis report for knowing suitable crop for growing, sell crops, organic certificate and claim crop insurance through mobile application by establishing a communication link between them.

2.2 Module description

The proposed android mobile application provides several services like IoT based soil nutrient and pH of irrigation water testing, promoting smart contract farming service between APMC and farm producers (farmers) by facilitating an interface to communicate with APMC to view the suitable crop to grow based on uploaded soil analysis report, agreement for purchasing the yields, insurance to crops and obtaining the organic certificate. Thus, the proposed smart farming based mobile application provides separate login modules for both APMC admin and Farmer. The mobile users are categorized based on the services namely Farmer Services and APMC Admin Services, facilities under each category are listed below.

2.2.1 Farmers services

In-order to extend the facilities to the farmer who is involved in organic farming, the mobile application is developed which will provide access to the farmer services on selecting user login option as farmer. Further the farmer can perform several activities in the application based on their choice. Each activity is explored in the following.

1. **IoT Enabled Sensor for Soil and Water [SWA] Analysis** – Soil and water are important factor that need to considered for effective growth of plants, so it is necessary to analyze these two viz. soil and water, for healthy and sustainable farming. Since the testing laboratories are located one per district and consume more time and delayed response from the testing laboratories for the reports. Hence the village farmers face difficulty in getting tested these parameters from the laboratory. So, the proposed IoT based smart farming mobile application helps in measuring the primary soil nutrients [NPK] and pH value of water using IoT enabled sensor implanted at different locations of the agricultural lands and water reservoirs. This automation will save the time and cost of the farmers. The block diagram of IoT enabled Soil NPK and pH of Irrigation Water testing is shown in below **Figure 2**.

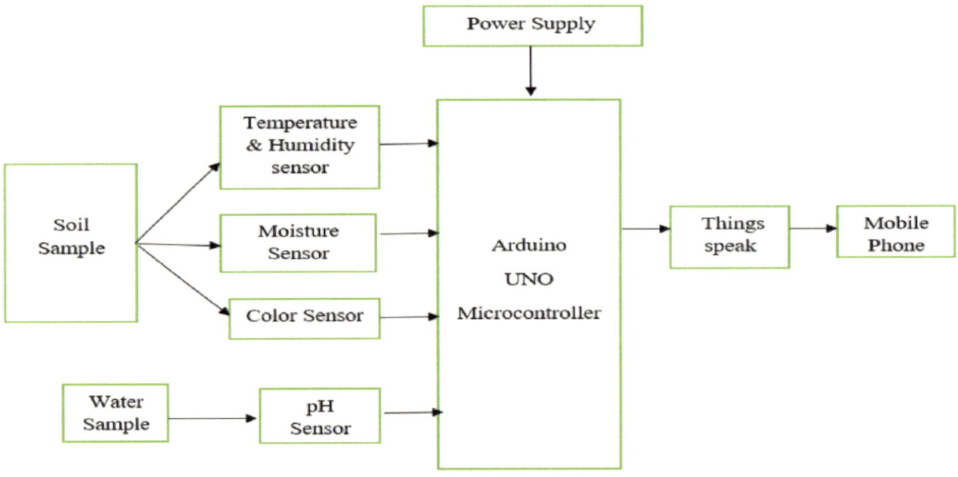

Figure 2.
Block diagram of soil NPK and pH of irrigation water analysis.

The prototype of IoT based soil NPK and pH of irrigation water testing makes use of hardware components like IoT enable Microcontroller, Ethernet shield, soil moisture, temperature and humidity sensor, Color Sensor, Water pH sensor and power supply.

2. **Communication with APMC** – This module helps farmers to register with APMC to access the facilities provided by the APMC authority. Services provided by APMC to the formers are depicted in the following.

   a. **About APMC** – This module provides information about the Agricultural Produce Market Committee (APMC).

   b. **Upload SWA Report** – It provides a facility for farmer to upload the soil and water analysis report of the agriculture land to the APMC admin for knowing the best and suitable crop for growing to get more yield.

   c. **View Crop Suggestion and Get Organic Certificate** – Admin will suggest the best and profitable crop to the farmers based on their uploaded SWA reports. Admin also offers an agreement clause and Organic certificate to the farmers which provides fair marketing for the farmers yield and also provide insurance facility for the crop.

   d. **Signing of Agreement** – It provides an option for the farmers to sign the agreement with the APMC. According to this agreement, the yield of the farmer was purchased by APMC at fixed price as mentioned in the agreement provided, if farmers agrees to grow crops that are suggested by APMC admin. The agreement clause also includes the Insurance for the crops in case yield loss due to occurrence of natural disaster like flood, droughts etc. Farmers will get the minimum amount per acre as mentioned in the agreement.

   e. **Sell Crops** – It gives an internet showcasing office/market to the farmer to offer their yields to the APMC at fixed cost.

2.2.2 APMC admin services

APMC is advertising board set up by state government in India to guarantee reasonable promoting for the farmer's yield. The services offered by the APMC admin are as follows.

1. View and Approve Farmer – It provides facility for admin to view the details of the farmer such as full name, mobile number and kisanId. If details are correct then admin will approve the registered farmer to provide access permission to use the APMC services.

2. View SWA Reports – This module provides facility for the admin to view the SWA reports uploaded by the farmers for further operation such as predicting and suggesting suitable crop to farmer.

3. Predict Crop - This module will help the admin to predict the best and suitable crop based on SWA reports using Machine Learning Algorithm. And also offers an agreement proposal to the farmers for ensuring the fair marketing to the farmers.

4. Buy Crop –APMC administrator can straightforwardly buy crop from ranchers utilizing this android application, it benefits both APMC administrator and ranchers to encourage advantageous reasonable promoting between them.

5. Insurance – It facilitates the admin to approve or reject the insurance claimed by the farmer by verifying the criteria.

In this section, the block diagram of the proposed model and description of its process flow is explained briefly. Detailed discussion and data flow diagrams of each module is explained in subsequent chapter. The soil nutrient and pH of irrigation water testing using IoT enabled sensor technology is explained.

3. System design

The process of system design helps in understanding functionality of the proposed system. The microscopic view of the functionalities of each module in the proposed work is described with the help of data flow diagram (DFD). Further to showcase the process, the case studies are considered and explored in the following sections.

The **Figure 3** shows the progression of information for login and enlistment movement procedure of the farmer. The farmer/user enters the login details. If the farmer does not have the login details then he/she needs to register to the mobile application. After successful registration of the farmer to the application the login details are saved then username and password is used for login into the application. Homepage of farmers mobile application consists of several modules and logout option is provided to logoff from the application.

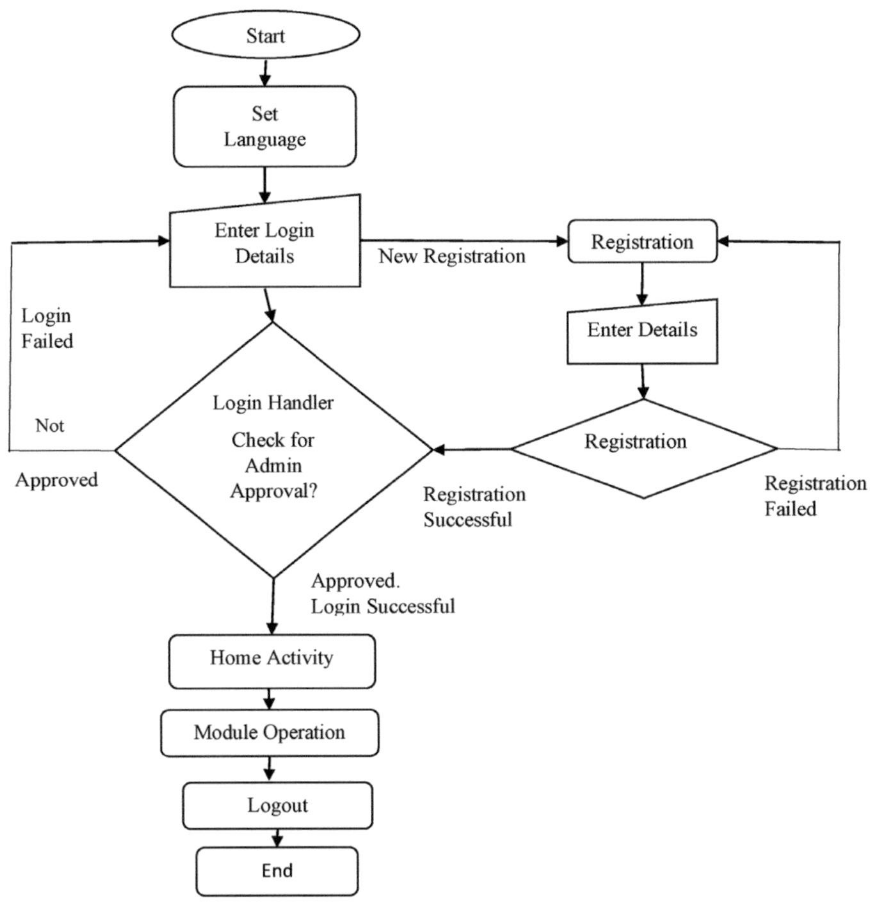

Figure 3.
DFD for login and registration activity of farmer.

The **Figure 4** shows the data flow diagrams of services present in the farmer homepage. The farmer needs to first choose upload SWA reading module because all other module operations depend upon upload SWA reading module. Then enter all the required details about the soil parameters and click upload button to upload values. Once after executing this module the rest all module operations get activated.

The admin should enter login details. If login details are correct then the admin login successfully. The **Figure 5** shows the data flow diagram of all the available services in the APMC admin homepage. The homepage of APMC admin has the logout button to logoff from the application.

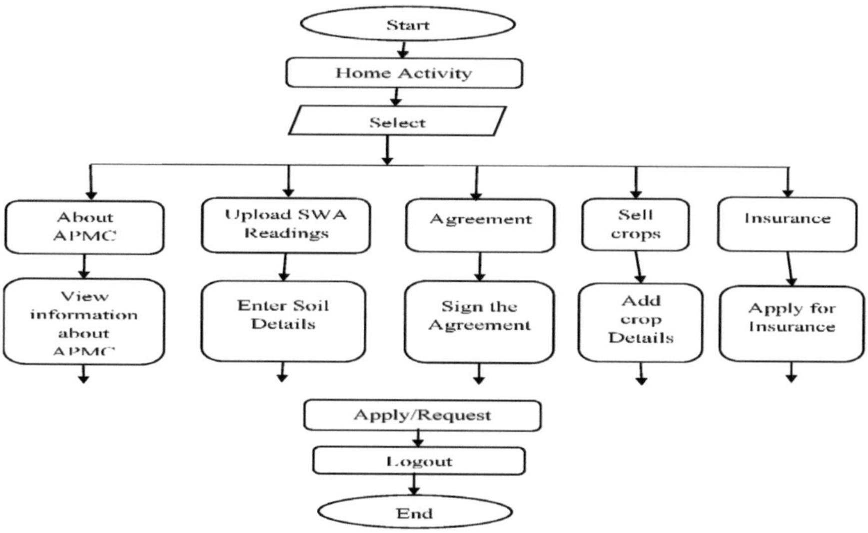

Figure 4.
DFD for applying/requesting process in farmer homepage.

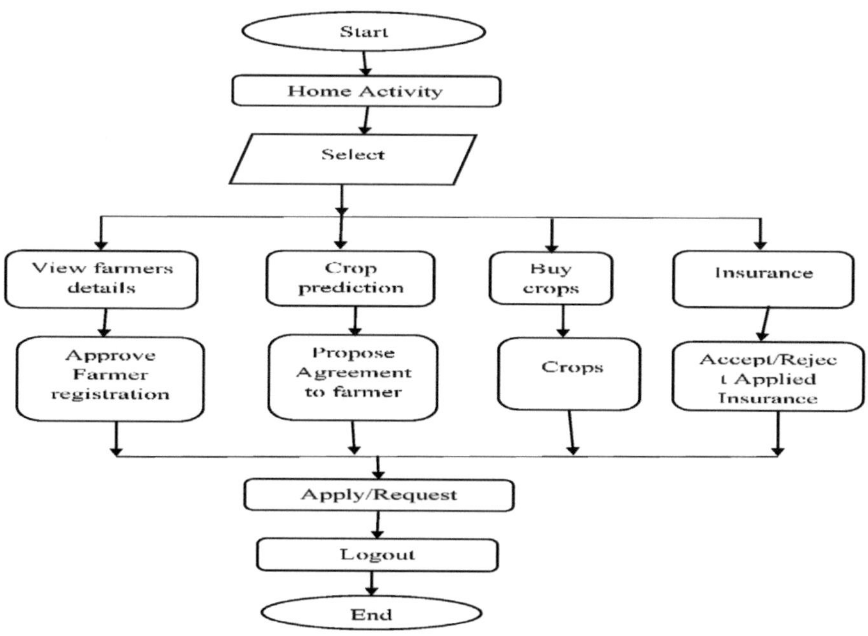

Figure 5.
DFD for applying/requesting process in admin homepage.

3.1 Modules dataflow

The design process helps to know how the developed mobile application is working in a particular way. There are basically two modules and each module associated with certain set of services.

3.1.1 Farmer module

The farmer can communicate with APMC officials to send SWA report collected from the IoT sensors, to get crop and organic farming suggestions for sustainable agriculture, direct marketing etc. The module description shows the services offered to the registered farmer of the mobile application and is depicted in the following.

3.1.1.1 Upload SWA readings

The government authorities have facilitated the agriculturist with Tablet PC/handheld devices. In-order utilize these handheld devices efficiently for agricultural activities, the proposed work developed a mobile application with two modules. The farmer module and APMC admin module. After successful login to the application, farmer can upload SWA reports by entering details of the soil and water analysis readings determined using IoT enabled sensor technology as shown in **Figure 6** for further processing.

3.1.1.2 View crop prediction and sign the agreement

The APMC admin offer an agreement proposal to the farmer along with suitable crop and fertilizer suggestion based on analysis done on the uploaded readings of soil parameter. Farmer can view the crop and fertilizer suggestion as show in **Figure 7** and have option to sign the agreement. This agreement states that the farmer must grow suggested crop and use the recommended fertilizer. It also provides insurance facility to the crops in case of yield loss due to occurrence of droughts and floods. It helps the farmers to grow crops with tension free because the purchasing price of the crop yield is fixed and mentioned in the agreement.

3.1.1.3 Sell crop

The farmer adds the information of the of the harvest which they need to sell, for example, crop name, amount and all out anticipated cost as shown in **Figure 8**. All these

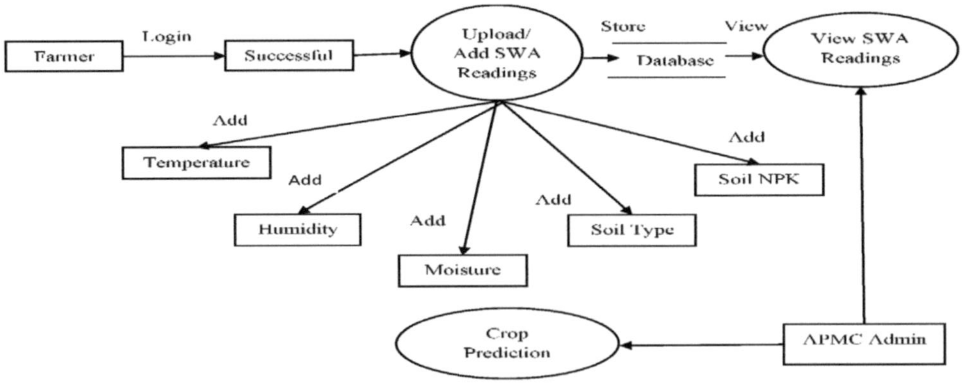

Figure 6.
Data flow for uploading SWA values.

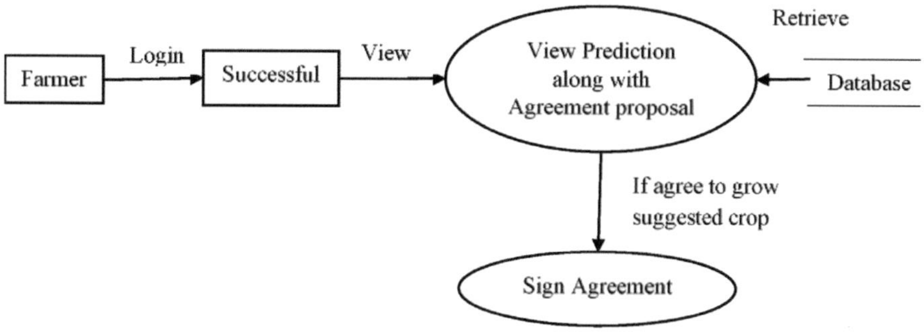

Figure 7.
Data flow for view crop prediction and signing of agreement.

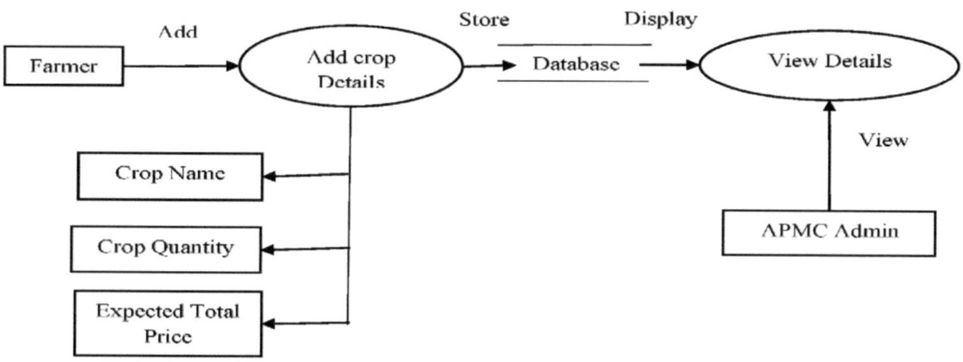

Figure 8.
Data flow for sell crop.

details of crop information are stored in the database so that APMC admin can retrieve and view crop details from the database to place an order for purchasing the yield.

3.1.1.4 Claim insurance

This service is given to farmer who signed the agreement offered by the APMC shop. Here farmer has facility to claim insurance through the mobile app in case of loss of crop yield due to occurrence of any unavoidable climatic condition such as droughts, floods etc. (**Figure 9**).

3.1.2 APMC admin module

Following module description shows the services offered by developed mobile application to the APMC Admin.

3.1.2.1 View and approve farmer details

The APMC admin has facility to view the farmer details such as name, kisanId and contact number of farmers along with status either approved or pending. The newly registered farmers have the status pending indicating that the admin has not yet approved the registered farmer. After verifying all the farmer details then admin has facility to approve the farmer to give access permission for the farmer to the mobile phone. The **Figure 10** shows flow diagram for verifying and approve farmer details. Thus, prevents the unauthorized person accessing the mobile app.

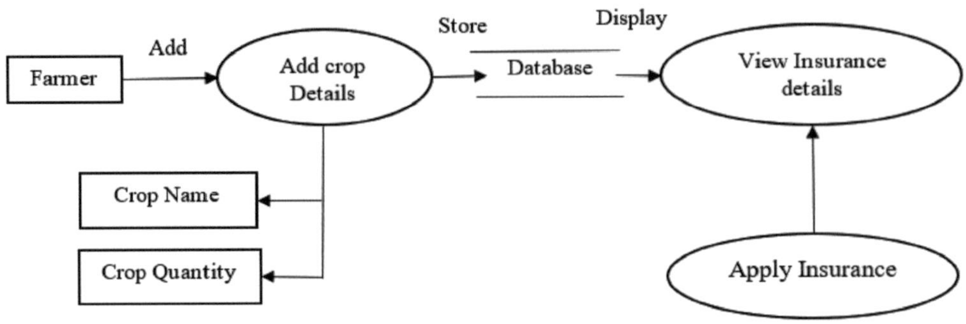

Figure 9.
Data flow for claiming insurance.

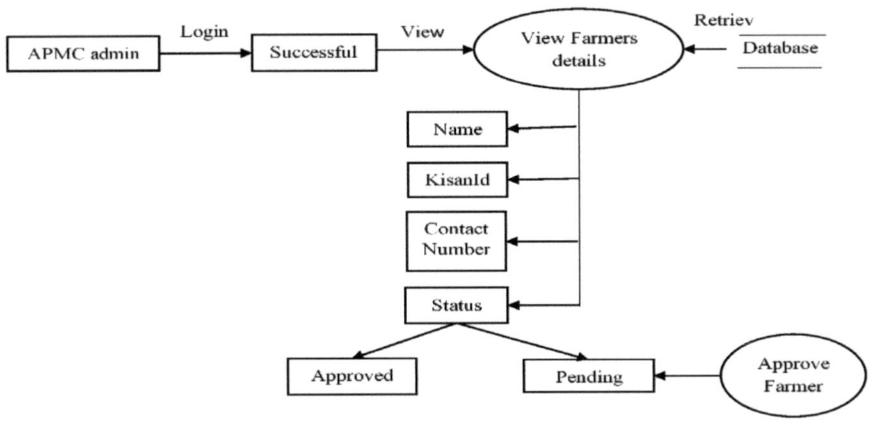

Figure 10.
Data flow for view and approve farmer details.

3.1.2.2 Crop prediction

The crop prediction is a process of predicting the suitable crop for location specific crop cultivation and appropriate fertilizer without degrading the quality of the soil for the sustainable agriculture. The five different steps which are followed in this module are shown in **Figure 11**.

1. Dataset collection

2. Data pre-processing

3. Data splitting into train and test sets

4. Fitting algorithm

5. Testing/Prediction

1. **Dataset Collection**

The dataset containing soil parameters are collected which includes moisture, temperature, humidity, soil type, NPK value, crop name and fertilizer name. The IoT enabled sensors data is stored in the matrix form. The grid size of the IoT enabled sensors from different locations of the agricultural land is 200 X 9. There are totally

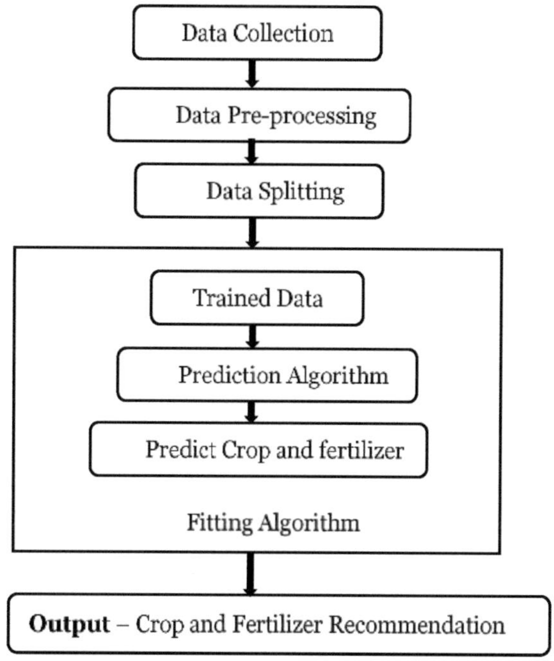

Figure 11.
Data flow for crop prediction.

200 rows and 9 columns. The first 3 columns contain data which were in string format i.e., NPK values and rest 6 columns contain data which are in numerical.

2. **Data pre-processing**

The information gathered from the IoT empowered sensors is in non-uniform and need preprocessing. In data pre-processing all the string data is converted into numerical data. In order to perform this, all the string features are converted into dummy variables which indirectly increased the column number, further the data is cleaned to remove null values. The dataset contains some categorical data and some continuous numeric data. This type of unstructured data cause problems to algorithm, hence the pre-processing task is performed standard feature scaling on all of the data to bring them into a common scale.

3. **Data Splitting**

The crop prediction module requires the supervised classifier such as support vector machine (SVM). In-order to train the supervised classifier, it is required to segment the accessible dataset into preparing and testing dataset. It is the way towards parting the dataset into preparing and testing information. In the proposed work the dataset is divided into an 80:20 ratio, the algorithm is trained using the training data and tested using test data to find the accuracy.

4. **Fitting algorithm**

The training the data file is to be carried out by loading the cpdata.csv file in order to separates features and labels that are done by applying fit module that is SVM (Support Vector Machine) algorithm used for classification, it works with managed learning method. In the event that the dataset comprise with the main highlights and names SVM works better. As SVM is primarily used for the binarization, binary classifier searches the hyper planes with possibility between positive and negative samples. The multi-SVM is used when the there are many classes from which the classification can happen successfully. There are various strategies offered, where a multi-class classifier is worked by blending the different parallel classifier and afterward used to prepare a SVM classifier in the choice tree root hub utilizing soil information.

5. **Prediction of Crop and Organic Certification**

The features extracted are used for prediction of crop and fertilizer for sustainable agriculture. The crop prediction is based on SWA (NPK values of Soil at different locations of the land and pH value of the water at different reservoirs) provided by the IoT enabled sensors. The parameters are validated and further the prediction for the crop is carried out. Further the authorities will issue the Organic certification for the suggested crop.

3.1.2.3 Buy crops

Once the farmer grow the organic product then farmer need to add the information about crop such as crop name, quantity and expected total price of the crop as shown in **Figure 12** which is stored in the database. If the direct buyer/APMC admin wish to purchase agriculture product/yield from the farmer they access the crop details from the database using mobile application to place an order for crop buy. The farmers can avoid the involvement of the mediators in the selling of crop and can earn maximum benefit.

3.1.2.4 Approve/reject insurance

In case of occurrences of any natural calamities/disaster such as landslides, flood, drought, fires etc. due to which farmers face loss in their organic farming yield. In-order to solve such issues the developed mobile application provides online insurance claiming facility for farmers. **Figure 13** shows the data flow diagram for approving/rejecting of the Insurance from the agency. They have to add the crop details to claim the insurance. Once the request to claim the insurance is sent to the APMC shop then the admin will verify the details of the crop with the existent data to check whether the farmer has grown the suggested crop or not. If the provided details are valid then the admin will approve the insurance else, he will reject the insurance.

In this section, the working flow of data is described using data flow diagrams. It also provides information about yield of every element and the procedure itself.

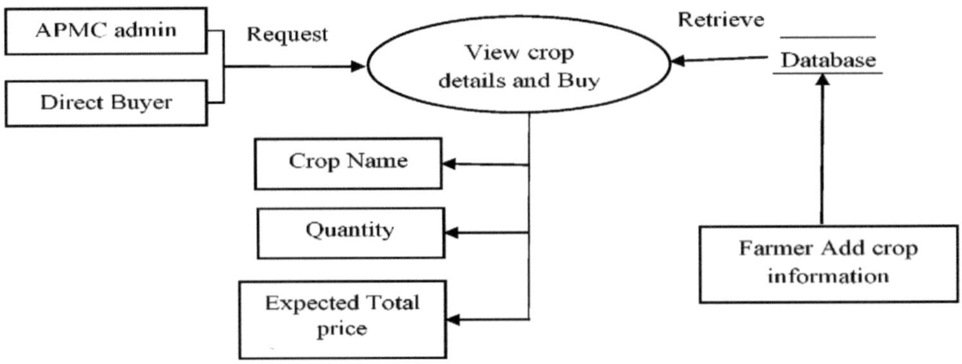

Figure 12.
Data flow for buying crop yield.

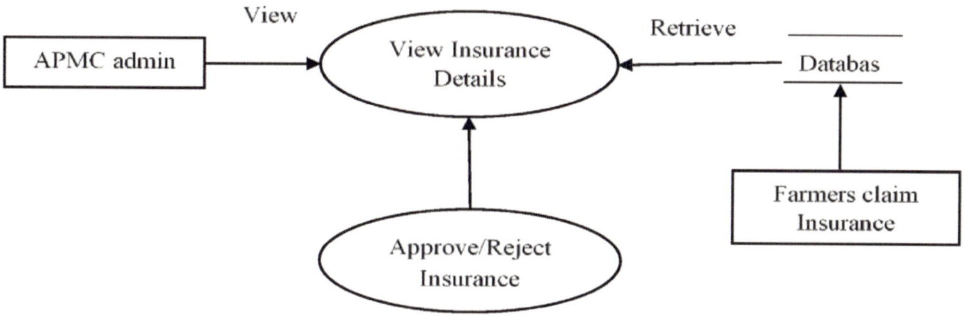

Figure 13.
Data flow for approve/reject insurance claim.

Objective of this system design deals with functionality each module in the system and also provides basic understanding of system characteristics.

4. Experimentation and results

The proposed organic farming model can be implemented in two phases, the first phase is building hardware kit using IoT enabled sensor technology to measure the soil macro nutrients [NPK] and pH of irrigation water. The second phase is developing advisory mobile application using Android studio to predict and suggest the suitable crop and appropriate fertilizer for organic farming to achieve sustainable agriculture. The experiment for measuring the soil parameter and pH of irrigation water has been conducted in fields for different soil samples. The results obtained from the sensor forms repository for further processing to predict crop and fertilizer using machine learning algorithm viz. SVM technique.

The **Figure 14** shows the experimental setup of hardware kit which consists of Arduino UNO microcontroller to control all the operation. The IoT enabled sensors namely soil moisture sensor to measure the moisture content of the soil, temperature and humidity sensor (DHT11) to measure the temperature and humidity of soil respectively, pH sensor for water pH value and color sensor is used to determine the soil NPK value.

Experimentation is carried out for six different parts of the Bagalkot district namely Hunugund, Kaladagi, Kerur, Bilagi, Kudalsangam and Mudhol of Karnataka State, India. The different lands of soil samples to measure the soil parameter such as moisture, temperature, humidity and NPK nutrient value of soil. The pH value of the

Figure 14.
Experimental setup of IoT enabled kit.

irrigation water is also determined at different locations of land including borewell water, rain water, river water etc. available in the reservoirs. The suitable pH value for the irrigation water for sustainable agriculture should lie between the value 5 to 7.

Experimentation is conducted as shown in **Figure 15** to measures the moisture, temperature and humidity content of soil in agriculture field which has grown green grams. The results are uploaded onto the thingsspeak software. This software helps to visualize and analyze live data by plotting graphs, the graph for moisture is shown in **Figure 16** and graph for temperature and humidity are as shown in **Figure 17**.

Figure 15.
Experimental setup of hardware kit to collect NPK values.

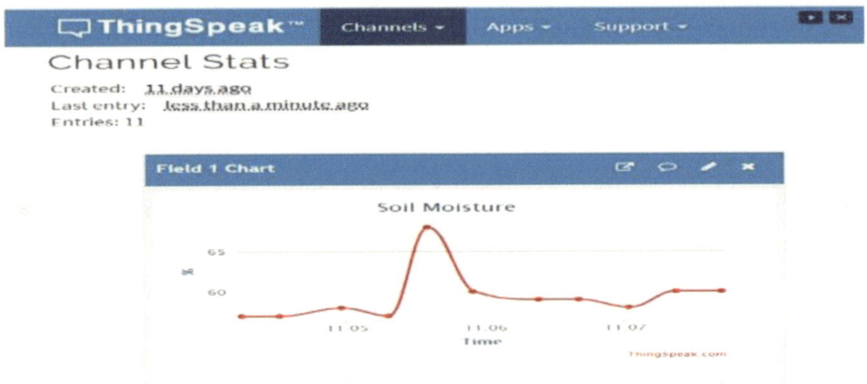

Figure 16.
Graph showing soil moisture.

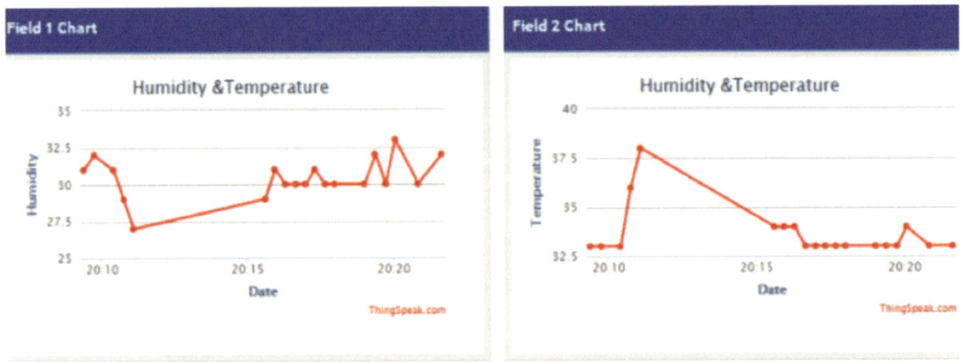

Figure 17.
Graph of humidity and temperature of soil.

Experimentation is additionally done to quantify these essential supplements and results are appeared in **Figures 18** and **19**.

The values which are retrieved from the sensor are transferred to Arduino from there to thingsspeak software and eventually on to the app. The result for different soil sample have been listed in the **Table 4**. With these measurements the sustainability is maintained and increased organic yield is obtained as depicted in the above **Table 1**.

Experimentation is also conducted to determine pH of irrigation water the LCD display value as shown in **Figure 20** shows the result of pH of water.

The Second step of implementation in the proposed methodology is developing the mobile application to assist the farmer in selecting location specific crop for cultivation and also provide appropriate fertilizer recommendation to the farmer. The developed mobile app can be used by both APMC admin and farmer as shown in **Figure 21** through which it establishes a communication link between them.

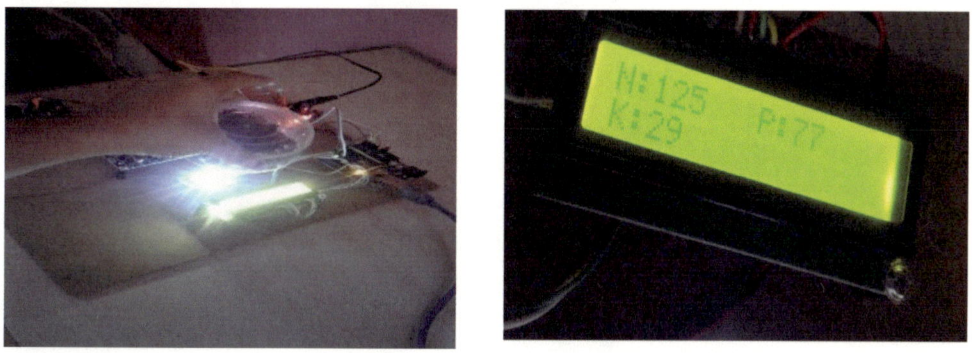

Figure 18.
LCD display output for soil sample 1.

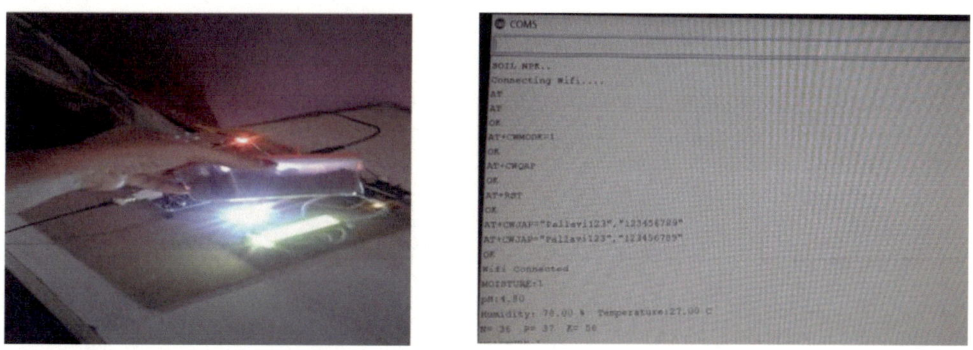

Figure 19.
Serial monitor output for soil sample 2.

Soil Sample No	Moisture	Temperature	Humidity	N	P	K
1	56	22	70	37	45	34
2	58	23	65	125	77	29
3	45	29	49	35	22	24
4	70	26	55	41	56	45
5	60	24	68	38	34	14

Table 4.
Results of sensor readings for different soil sample.

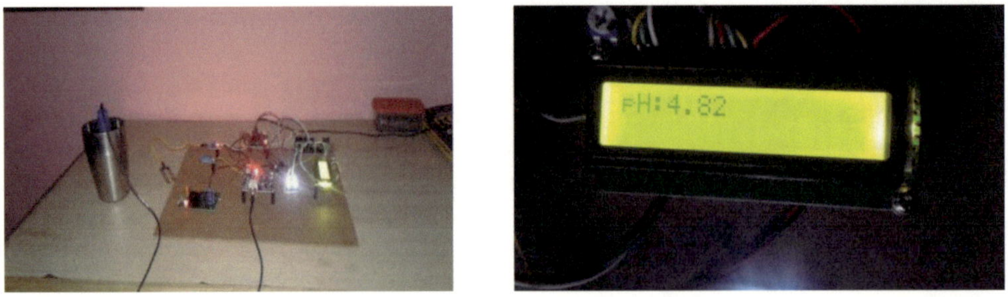

Figure 20.
LCD display value of pH of irrigation water.

Figure 21.
Mobile application - direct marketing.

4.1 Farmer registration and login

The application should be installed in the smart phone and farmer need to register to the application as shown in **Figure 22** after successful registration the details

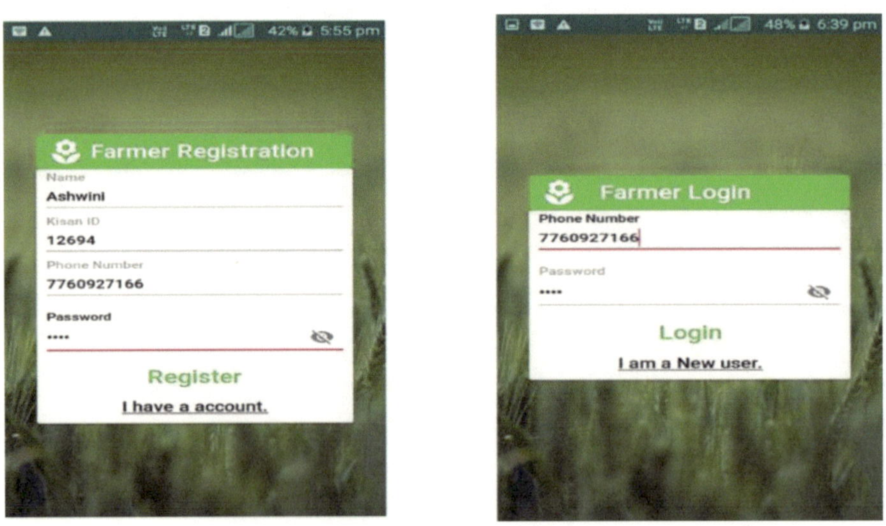

Figure 22.
Snapshot of registration and login page of farmer.

will be sent to admin for approval. Once the admin approved the farmer details then the farmer can able to login to application using username and password as shown in **Figure 22**.

4.2 APMC admin login and home page

Agricultural Produce Market Committee (APMC) is a state government body which ensures safeguard to the farmers from exploitation by large retailers. The admin can login to the application using username and password as shown in **Figure 23** after successful login the admin can access the home page.

4.3 View and approve farmer details

The admin has the facility to view the farmer details as shown in **Figure 24** the status pending shows that the farmer is newly registered and admin will approve the farmer by just clicking on the checkbox button. The status approved indicates that the farmer is an existing user of the application. Only approved formers can successfully login and access the homepage services as shown in **Figure 25**.

4.4 Upload readings

In Upload soil and water analysis (SWA) readings the details such as moisture, temperature, humidity and NPK value are filled as shown in **Figure 26** after clicking the upload button the details will be uploaded to the database for further processing. The APMC admin can use this information for crop and fertilizer prediction as shown in **Figure 27**.

4.5 View prediction and sign agreement

The farmer can view the suitable crop and appropriate fertilizer as shown in **Figure 27** suggested by the admin. Along with this the admin offers an agreement proposal to the farmer. According to this agreement the farmer has to grow only those crops that are suggested by the admin, the purchasing price of the crop yield is fixed in the agreement before the farmer starts growing it and it also provides crop insurance

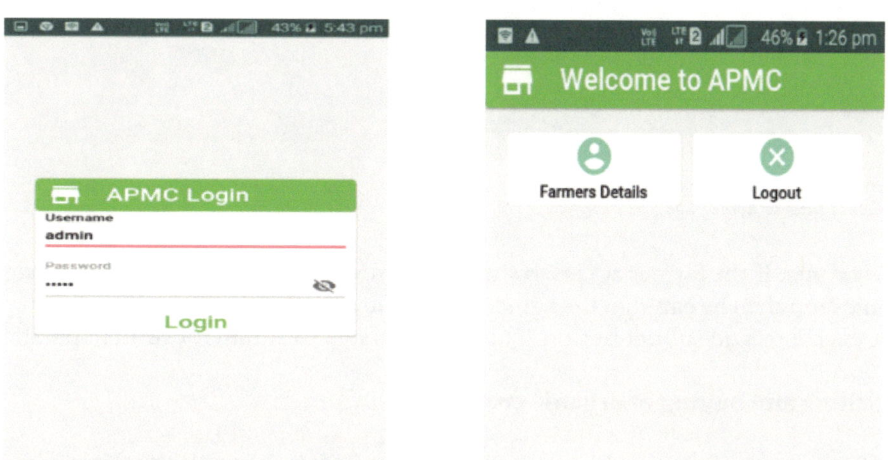

Figure 23.
Snapshot of admin login and home page.

Figure 24.
Snapshot of view and approve farmer.

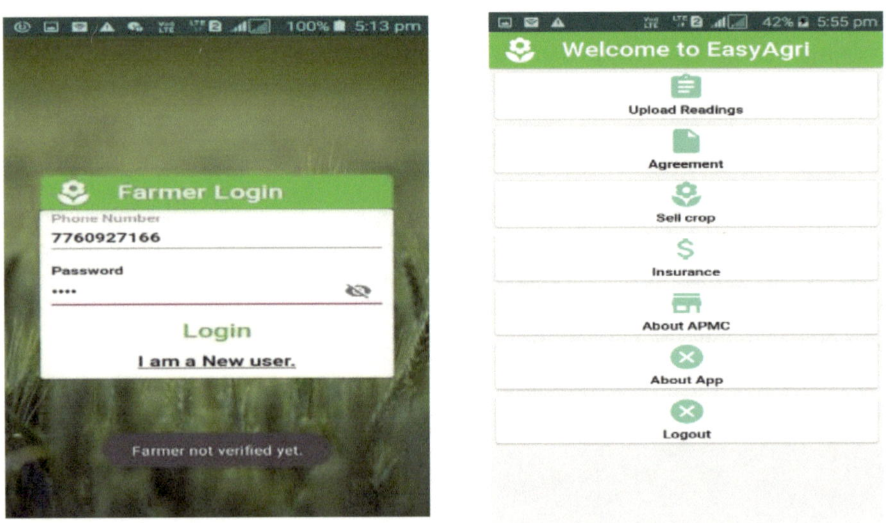

Figure 25.
Snapshot of farmer home page.

to the farmer. If the farmer agrees to stated agreement criteria to grow the suggested organic crop then he can sign the agreement as shown in **Figure 28** by just clicking I agree checkbox and submit button. The request is sent to admin for future operation.

4.6 Selling and buying of organic crops

Whenever the farmer wishes to sell the crop yield, he/she fills the details as shown in **Figure 29** the crop details will be stored in the database and when the admin login to app to buy the crops then these stored details will be retrieved and displayed to admin.

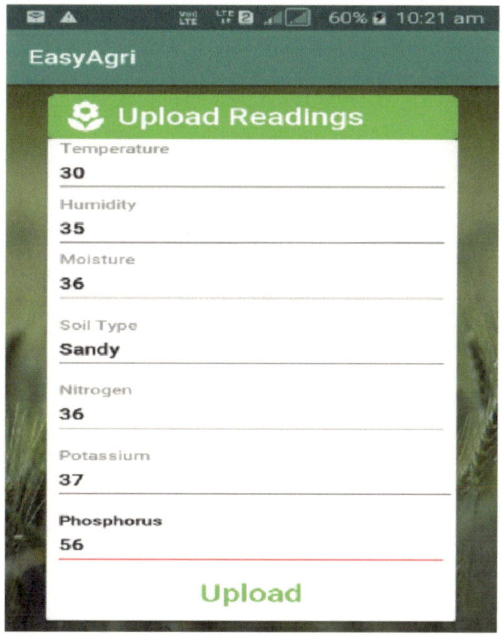

Figure 26.
Snapshot of uploading SWA readings.

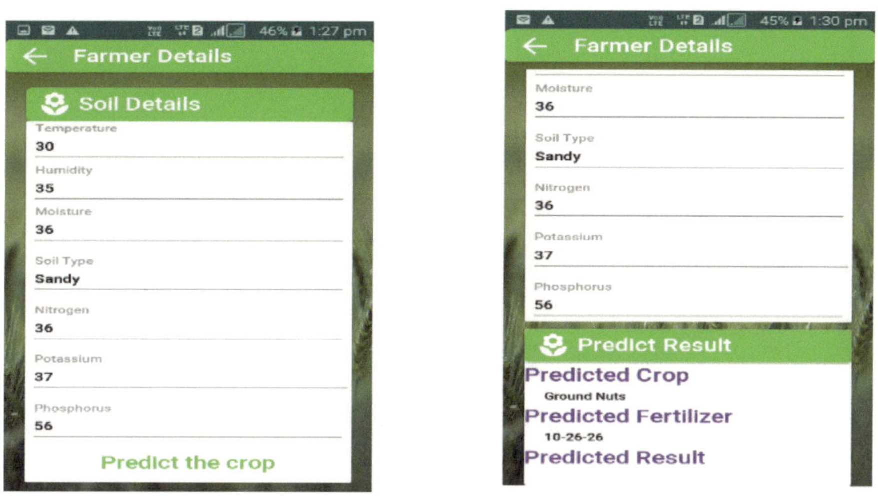

Figure 27.
Snapshots of organic crop prediction.

4.7 Insurance

Insurance is a means of protection from financial loss. The agreement provides the crop insurance to the farmer, to avail this facility farmer must have signed the agreement. It is given to the farmer whose yield is lost due to occurrences of unavoidable climatic conditions such as droughts, floods etc. to save the farmers from financial loss. Using this mobile app farmer can claim the insurance by providing the necessary details as shown in the **Figure 30**. The request is sent to the admin for approval, on verifying the details admin has the facility to approve or reject the insurance claim as shown in **Figure 30**.

This section shows the experimental results of complete implementation of the proposed method. The experiment is conducted in agriculture fields to measure real time values of soil parameter such as moisture, temperature, humidity and NPK nutrients

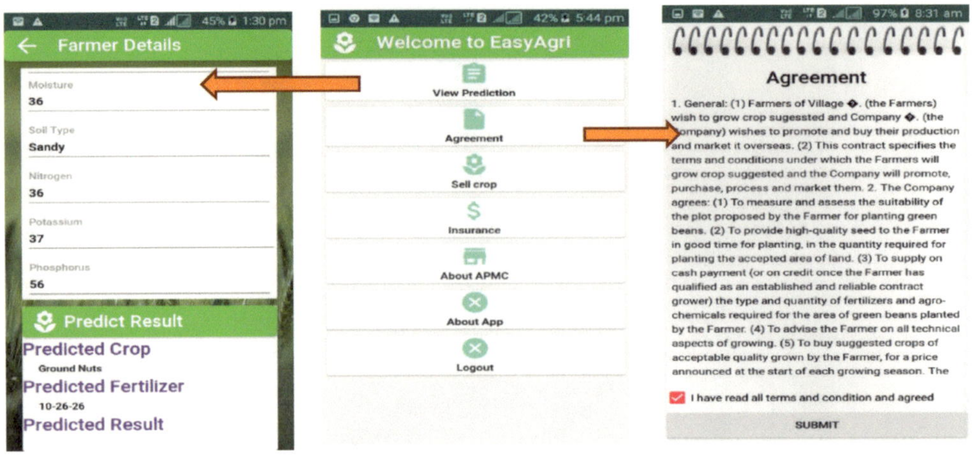

Figure 28.
Snapshot of view prediction and signing of agreement.

Figure 29.
Snapshot of selling and buying of organic crops.

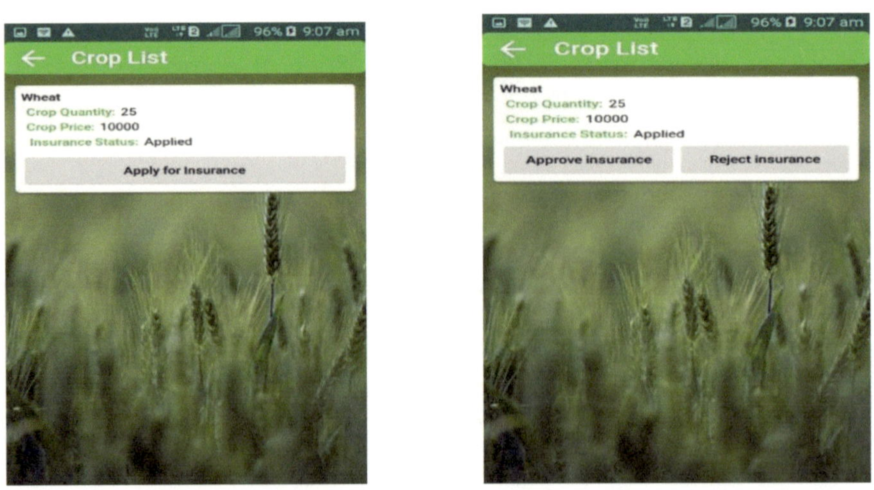

Figure 30.
Snapshot of insurance process.

value and listed the read value in the **Table 3**. Further these values are used for predicting suitable crop and appropriate fertilizer. The yield of the organic farm is monitored with the help of Mobile and IoT enabled technology for the duration of one year and tabulated the observations, which indicated the production of the yield is increased and the soil fertility is also improved. The proposed methodology obtained the promising results for sustainable agriculture using soil and water nutrients and farmers get direct marketing to their increased organic products at APMC as per the agreement.

5. Conclusion

The proposed system of an organic farming for sustainable agriculture is an advisory mobile application which helps the farmers to self-test their field parameters for generating quick soil and water analysis report and developed android mobile application predicts and suggest the suitable crop to farmers along with agreement proposal to meet the sustainable agriculture practice. It helps to maintain and improve the soil fertility creating ecologically sustainable environment. Thus, the IoT enabled sensor kit helps the farmers to get the soil testing services at the doorstep. It also provides awareness and usage of tools and technology. Hence overall mobile app is used to automate the agricultural process.

6. Future enhancement

The work can be expanded to implement on large fields and skill-based trainings on usage and monitoring of sensor module given to the farmers which helps them to easy the farming practice. Further the developed organic farming based mobile application can also be expanded to implement the government initiative scheme for "One District One Crop" functionality to augment crop diversity.

Author details

Sanjeevakumar M. Hatture[1], Pallavi V. Yankati[1*], Rashmi Saini[2] and Rashmi P. Karchi[3]

1 Department of Computer Science and Engineering, Basaveshwar Engineering College (Autonomous), Bagalkot, Karnataka, India

2 Department of Computer Science and Engineering, G.B. Pant Institute of Engineering and Technology, Pauri Garhwal, Uttarakhand, India

3 Department of Computer Science and Statistics, Vishwachetan College, Bagalkot, Karnataka, India

*Address all correspondence to: pallavivijay3575@gmail.com

References

[1] Nancy Bundan, Kim-Mey Chew, "IoT Soil Moisture Monitoring and Irrigation System Development", Proceedings of the 9th International Conference on Software and Computer Application, February 2020, pp. 247-252

[2] Suvarna Nandyal, Priya S Khamitkar, "Agridrone: Automation of Agriculture using IoT", June 2019, International Journal of Innovative Science and Research Technology, Volume 4, Issue 6.

[3] Sanjeevakumar M. Hatture, Pallavi V. Yankati (2021), IoT-Based Smart Farming Application for Sustainable Agriculture. In: Tuba M., Akashe S., Joshi A. (eds) ICT Systems and Sustainability. Advances in Intelligent Systems and Computing, vol. 1270. Springer, Singapore. https://doi.org/10.1007/978-981-15-8289-9_56.

[4] Sujatha Anand, Silviya Catherine, "Monitoring of Soil Nutrients Using IOT For Optimizing the Use of Fertilizers", April 2019, International Journal of Science, Engineering and Technology Research (IJSETR), Volume 8, Issue 4.

[5] Ha Anh Minh Tran, Ha Quang Thinh Ngo, "Design of Green Agriculture System Using Internet of Things and Image Processing Techniques", 2018 IEEE 4th International Conference on Green Technology and Sustainable Development (GTSD), 2018

[6] Sowmiya E, S. Sivaranjani, "Smart System Monitoring On Soil Using Internet Of Things (IOT)", Feb -2017, International Research Journal of Engineering and Technology (IRJET), Volume: 04 Issue: 02.

[7] Anggy Pradiftha Junfitrhana, Muhammad Labib Langlangbuana, "Developing Potential Agriculture Land Detector for Determine Suitable Plant Using Raspberry-Pi", 2017, International Conference on Computing, Engineering, and Design (ICCED)

[8] Sreeram.K, Suresh Kumar. R, "Smart Farming—A Prototype for Field Monitoring and Automation in Agriculture", International Conference on Wireless Communications, Signal Processing and Networking (WiSPNET), 22-24 March 2017, 10.1109/WiSPNET.2017.8300148

[9] Ramya Venkatesan and Anandhi Tamilvanan, "A Sustainable Agricultural System Using IoT", International Conference on Communication and Signal Processing, April 6-8, 2017, India,

[10] P.R. Harshani, and T. Umamaheswari, "Effective Crop Productivity and Nutrient Level Monitoring", International Conference on Soft-computing and Network Security (ICSNS), 2018.

[11] Snigdha Sen, Madhu B, "Smart Agriculture: A bliss to Farmers", International Journal of Engineering Sciences & Research Technology, ISSN: 2277-9655, April, 2017.

[12] K. A. Patil, N. R. Kale, "A Model for Smart Agriculture Using IoT", International Conference on Global Trends in Signal Processing, Information Computing and Communication, 2016.

[13] Kavianand G, Nivas V M et al, "Smart Drip Irrigation System for sustainable Agriculture", 2016 IEEE International Conference on Technological Innovations in ICT For Agriculture and Rural Development (TIAR 2016).

[14] Amrutha A, Lekha R, A Sreedevi, "Automatic Soil Nutrient Detection and Fertilizer Dispensary System", IEEE International Conference on Robotics: Current Trends and Future Challenges (RCTFC), 2016.

[15] Prachi Sharma, D.V. Padole, "Design And Implementation Soil Analyser Using IoT", 2017 International Conference on Innovations in Information, Embedded and Communication Systems (ICIIECS)

[16] Sriram Rakshith. K, Dr. Deepak. G, Rajesh M, Sudharshan K S, Vasanth S, Harish Kumar N, "A Survey on Crop Prediction using Machine Learning Approach", Apr 2019, International Journal for Research in Applied Science & Engineering Technology (IJRASET), Vol 7, Issue IV

[17] S. Bhanumathi, M. Vineeth et al, "Crop Yield Prediction and Efficient use of Fertilizers", IEEE International Conference on Communication and Signal Processing, April 4-6, 2019, India, 2019

[18] Sanjeevakumar M. Hatture, Susen Naik, "Agro Guardian: A Smart Agriculture Framework for Precision Farming", Book Chapter in Modern Techniques for Agricultural Disease Management and Crop Yield Prediction, Vol. 8, pp. 179-202, IGI Global, 2019.

[19] Thayakaran Selvanayagam, Suganya S et al, "Agro-Genius: Crop Prediction using Machine Learning", International Journal of Innovative Science and Research Technology, Volume 4, Issue 10, October – 2019

[20] Nitin Singh; Saurabh Chaturvedi, "Weather Forecasting Using Machine Learning Algorithm", International Conference on Signal Processing and Communication (ICSC), NOIDA, India, 7-9 March 2019, DOI: 10.1109/ICSC45622.2019.8938211

[21] C.P. Wickramasinghe, P.L.N. Lakshitha et al, "Smart Crop and Fertilizer Prediction System", International Conference on Advancements in Computing (ICAC), Malabe, Sri Lanka, 5-7 Dec. 2019, DOI:10.1109/ICAC49085.2019.9103422

[22] Sk Al Zaminur Rahman, Kaushik Chandra Mitra et al, "Soil Classification using Machine Learning Methods and Crop Suggestion Based on Soil Series", 2018 21st International Conference of Computer and Information Technology (ICCIT), 21-23 December, 2018

[23] P. S. Vijayabaskar, Sreemathi. R et al, "Crop Prediction Using Predictive Analytics", IEEE International Conference on Computation of Power, Energy, Information and Communication (ICCPEIC), 2017.

[24] D.S. Zingade, Omkar Buchade et al, "Machine Learning based Crop Prediction System Using Multi-Linear Regression", International Journal of Emerging Technology and Computer Science, Volume: 3 Issue: 2 April – 2018

[25] P. Priya, U. Muthaiah et al, "Predicting Yield of the Crop Using Machine Learning Algorithm", International Journal of Engineering Sciences & Research Technology, April, 2018, DOI: 10.5281/zenodo.1212821

[26] Sanjeevakumar M. Hatture, Pallavi V. Yankati, "Smart Farming for Sustainable Agriculture using Water and Soil Nutrients", Annual Technical Volume of the Agricultural Engineering Division (ATV AGDB III 2020) of the IEI on "Agricultural Automation and Engineering Standards", pp. 62-67, December 2020

Chapter 4

The Role of Organic Fertilizers in Transition to Sustainable Agriculture in the MENA Region

Helen Avery

Abstract

Organic fertilizers can serve as an element of transitions to sustainable low-input agriculture in semi-arid regions of the MENA region. They play a key role in supporting soil biota and soil fertility. Yield improvements, availability and relatively low costs make organic fertilizers an attractive alternative for farmers. In semi-arid regions, important considerations are improved soil quality, which in turn affects soil water retention, while better root development helps crops resist heat and water stress. Organic fertilizers thus support climate adaptation and regional food security. Soil quality is crucial for carbon sequestration, at the same time that increased nutrient retention reduces impacts of agricultural runoff on groundwater and water bodies. Factors that impede the generalised use of organic fertilizers include lack of expertise, subsidy structures, constraints of the wider food and agricultural systems, and difficulties in transitioning from conventional agriculture. Such obstacles are aggravated in countries affected by security issues, financial volatility or restrictions in access to market. Against the background of both general and local constraints, the chapter examines possible pathways to benefit from organic fertilizers, in particular synergies with other sustainable agricultural practices, as well as improved access to expertise.

Keywords: organic fertilizers, sustainable agriculture, transition pathways, smallholder farmers, semi-arid regions, low-input agriculture, soil health, soil carbon, GHG emissions, conservation agriculture, water management, climate adaptation and mitigation

1. Introduction

Organic fertilizers are a highly diverse family of products used in agriculture for soil improvement and to provide nutrients. Their characteristics and benefits will depend on their origin and processing, as on how they are used or combined in particular contexts [1–5]. The main common denominator is therefore that organic fertilizers provide a sustainable option to avoid the negative impacts of chemical fertilizers for long term soil fertility [6], decrease vulnerability to climate stress and weather variability, while reducing the impacts of agriculture on the environment [7, 8].

The term 'organic fertilizers' refers to a very wide range of products, as do the terms chemical, inorganic or synthetic fertilizers. It is therefore exceedingly difficult to make sweeping generalisations concerning the respective benefits or characteristics of these types of fertilizers. The task becomes all the more challenging, since outcomes will depend on numerous factors. These include how the fertilizer matches soil characteristics, crops, climatic and topographical questions, landscape characteristics, but also irrigation and tilling practices, time and manner of application of the fertilizer, as well as details concerning source and manner of producing the fertilizer. Undesirable effects may result from inappropriate fertilizer production processes, and the presence of metals and other contaminants in source materials is a major concern [9, 10]. There are also challenges linked to the overall or local availability of source materials.

Using organic matter to improve soils is not only related to fertility, but also to effects on physical, chemical and biological soil properties, including aeration, permeability, water-holding capacity and nutrient preserving capacity [11]. Benefits will depend on the exact type of organic fertilizer used, as well as on soil characteristics [7, 11]. Organic fertilizers can be used alone, or in combination with other fertilizers. For instance, a study under experimental conditions suggests that under deficit irrigation conditions, a combination of chemical fertilizer with vermicompost produced better results than chemical fertilizer alone [12]. The use of organic fertilizers appears particularly interesting in conditions of stress and weather variability, while a tailored combination with micro-nutrients suitable for crop and soil enhances yields (see e.g., Parmar et al. [13]). However, much of the literature on fertilizers reduces outcome to the question of crop yield rather than resilience, and more specifically short-term gains in crop yield under normal circumstances.

The use of synthetic fertilizers was generalised as part of the so-called green revolution [14, 15], which stood for a vision of modernising agriculture through use of agricultural machinery, synthetic fertilizers, pesticides, and systematic improvement of crop varieties. The ambition was to dramatically increase food production, and thereby alleviate hunger globally, so the focus on short term crop yield is therefore not surprising. The vision of the green revolution was also very much part of an industrial paradigm, with a simplified vision of agriculture as resembling other industrial production processes, with a flow consisting of input and output, controlled process, and output, where success was measured in production units. Today, however, we have come to a realisation that this oversimplification brought with it a very high cost to the environment, human health, as well as a degradation of planetary conditions necessary for food production in the long term. Crop yields remain important, of course, but there are other implications of our choice of agricultural practices that equally need to be considered. While much of agronomical research investigates linear correlations between a small set of isolated factors under relatively stable conditions, Hou et al. [16] argue for the need to consider soil health holistically, dynamically and from an interdisciplinary perspective.

Besides the narrow focus on productivity, the industrial paradigm within which agriculture was placed has tended to favour a comparatively linear and mechanistic understanding, while disregarding the complexity of ecosystems below ground, above ground, and in water bodies. Soil exchanges gases and chemical substances with air, and aerosols from erosion, burning and vegetation affect cloud formation, precipitation and greenhouse effects [17–19]. Also, as farmers have always known, weather is highly unpredictable, and far from the controlled conditions that industrial production supposes. In view of current rapid climate change [20], farmers are facing increasing weather variability, a greater number of extreme events, and a greater extent of uncertainty with respect to future developments [21, 22]. The use of organic fertilizers alone is not sufficient to address these challenges but can, in combination with other sustainable agricultural practices, constitute an important ingredient in farmers' climate adaptation and mitigation strategies.

2. Agriculture in the Middle East and North Africa

Soil types, crops and trade patterns vary considerably across the Middle East and North Africa (MENA) region [23], but all countries are affected by water scarcity. The region comprises arid, semi-arid and hyper-arid areas, but even comparatively water-rich countries are affected by severe water stress [24], caused in part by economic incentives to cultivate water-intensive crops. Crop choice therefore plays an important role [25]. The water crisis is aggravated by deterioration of water quality caused by pesticides and nutrient runoff [26, 27], while groundwater is impacted by leaching and excessive pumping [28, 29]. Rural flight and decline of rural populations in several countries, such as Iran and Turkey [30] can reflect reduced need for labour due to mechanisation but may also reflect insecure livelihoods and difficult conditions of farmers [31, 32], while rural populations are also affected by displacement caused by disasters related to extreme weather, including forest fires, flooding and crop failure. The region is heavily dependent on imports of cereals. Both price fluctuations and transitions away from hydrocarbons globally will lead to decline in hydrocarbons exports on which many states of the region depend, affecting their ability to ensure food security through imports [23]. However, vested interests in exploiting hydrocarbons for the production of petrochemicals for agricultural use, as well as the existence of major phosphate deposits are likely to influence national economic diversification policies.

Large parts of the Middle East and North Africa are affected by protracted conflicts, internally displaced populations, and high volatility [33, 34]. Political and economic crises are affecting access to food, clean water and energy for large population groups [35], while agriculture is impacted by rising costs of fertilizers, pesticides, fuel and machinery, combined with disruptions to infrastructure and processing, storage and distribution systems for agricultural produce. These challenges will increasingly be aggravated by climate change [36–41] and environmental degradation. Consequently, resilient food systems and food security will become issues of major concern for the region [42, 43], highlighting the question of climate adaptation strategies for farmers [31, 44–46].

Research on organic fertilizers in the MENA region from an environmental perspective is as yet relatively limited. Thus, a Scopus search on October 14, 2021, with the search term 'organic fertilizers' yielded 517 articles and reviews in English concerning agricultural sciences in the MENA region for the period 2017–2021, compared to 6558 worldwide for the same period. Publications in this field were dominated by Iran, Iraq, Egypt and Turkey (92%). Only 102 (20%) of the 517 MENA publications related to environmental or earth and planetary sciences. Within these 102, a mere 5 directly dealt with water-related issues, (including keywords such as irrigation, water quality, water stress, arid regions or groundwater), and none of the overall 517 publications on organic fertilizers mentioned climate adaptation or mitigation. In view of the interrelated urgent challenges that climate change and food security pose for the region, I will therefore draw on the international literature, to situate the use of organic fertilizers with respect to these challenges.

3. Environmental impacts of agriculture

Climate and environmental impacts of fertilizer use and soil management practices include not only emissions and pollution from production of fertilizer [47], but also those linked to the mechanised and chemical-intensive agricultural production systems they are associated with, impacts of nutrient runoff and chemicals [48, 49] on receiving water bodies, as well as impacts connected to food processing, storage, transport and waste. Effects on the world's oceans are concerning. Unsustainable land use poses a threat for climate and biodiversity [20, 36, 50]. Agricultural land use and soil management practices are from a climate and environmental perspective of relevance for carbon storage [51], but also with respect to nutrient runoff, and persistent chemicals, and to emissions of N_2O and CH_4 [52]. According to the IPCC, the use of fertilizers has increased nine-fold since 1961 [53], and soil management accounts for half of greenhouse gas (GHG) emissions of the agricultural sector [54].

3.1 IPCC estimates of climate impacts and mitigation potentials

No global data are available specifically for agricultural CO_2 emissions, and there is considerable uncertainty concerning net balance of CO_2 land-atmosphere exchanges. However, land is an overall carbon sink, with a net land-atmosphere flux from response of vegetation and soils of −6 ± 3.7 GtCo2yr (averages for 2007–2016). The capacity of land to act as a carbon sink is expected to decrease as an effect of global warming. The major impacts of agricultural land use (food, fibre and biomass production) on CO_2 (5.2 ± 2.6 GtCo2yr) are connected to deforestation, drainage of soils and biomass burning rather than to the net flux balance directly caused by different fertilization practices. Numbers regarding CO_2 emissions from land use can be compared to net global anthropogenic CO_2 emissions, which are estimated at 39.1 ± 3.2 GtCo2yr. In addition to land use impacts, agriculture causes CO_2 emissions in the order of 2.6–5.2 GtCo2yr through activities in the global food system, including grain drying, international trade, synthesis of inorganic fertilizers, heating in greenhouses, manufacturing of farm inputs, and agri-food processing [55].

Agricultural land use directly represents 40% (4.0 ± 1.2 GtCo2eq yr) of total net global anthropogenic CH_4 emissions, and represents 79% (2.2 ± 0.7 GtCo2eq yr) of total net global N_2O emissions. CH_4 emissions are mainly caused by ruminants and rice cultivation. Half of N_2O emissions are caused by livestock, and the rest mainly by N fertilization (including inefficiencies). Total average net global GHG emissions (CO_2, CH_4 and N_2O) for all sectors 2007-2016 are estimated at 52.0 ± 4.5 GtCo2eq yr, of which agriculture directly contributes with 17-22% (not including impacts of agriculture on land available for forests), or 21-37% (including agricultural land expansion and other contributions of the food system) [55]. Importantly, agricultural soil carbon stock change is not included in these statistics. Irrigation and agricultural land management contribute to making forests vulnerable to fires, while desertification [37] amplifies global warming through release of CO_2, but such emissions as well as impacts from runoff on net fluxes from wetlands, water bodies and oceans are not included in the above figures.

Although net GHG emissions are often converted to CO_2 equivalents for accounting purposes, different gases remain in the atmosphere for different periods of time and will consequently have different impacts on the progression of global warming. The specific proportions of GHG will affect the likelihood of crossing critical thresholds and tipping points, setting off cascades (cf. Lenton et al. [56]) with ecosystem collapse and mass extinctions, while driving biophysical processes that further aggravate the dynamics. Effects of mitigation measures also have varying timelines.

The creation of reactive N in agriculture has significant environmental impacts [57], and excessive application of nitrogen can increase nitrous oxide emissions without improving crop yields [54]. On average, only 50% of N is used, but in countries with heavy N fertilization the efficiency can be much lower, and the potential for mitigation therefore increases [7, 36]. Use of fertilizer is responsible for more than 80% of N_2O emissions increase since the preindustrial era [58]. Ruminant livestock is the overall main source of CH_4 from agricultural practices [55, 59], and among organic fertilizers cattle manure has therefore been widely studied. Rice cultivation makes the greatest contribution to CH_4 emissions from agricultural soils [60]. Both water logging and soil compaction also contribute to CH_4 emissions [61].

4. Climate mitigation potentials in agriculture

In view of the imminent threat to planetary life systems posed by climate change [20, 56, 62], research has in recent years accelerated on potentials for carbon offsetting and impacts on GHG emissions of different land use and management systems [63–67], as well as with respect to climate adaptation [68] and food security [69, 70]. Large areas of the MENA-region are exposed to desertification, including relatively water rich countries. For instance, at least half of Turkey is affected [37]. Desertification amplifies global warming through the release of CO_2 linked with the decrease in vegetation cover, GHG fluxes, sand and dust. In dry areas, net carbon uptake is about 27% lower than elsewhere, reducing the capacity of land to act as a carbon sink. A rise in temperatures accelerates decomposition, at the same time that moisture is insufficient for plant productivity. Further SOC is lost due to soil erosion. An estimated 241–470 GtC is stored in the top 1 m of dryland soils [37]. In 2011, semi-arid ecosystems in the southern hemisphere represented half of the global net carbon sink [37].

Integrated sustainable practices are essential for climate adaptation, but estimates with respect to mitigation potentials vary. The chapter on interlinkages in the IPCC special report on climate change, desertification, land degradation, sustainable land management, food security, and greenhouse gas fluxes in terrestrial ecosystems [8] considers technical and economic feasibility of possible mitigation measures, as well as impacts on livelihoods and human health. Some measures that specifically concern cropland and soil management are summarized in **Table 1**.

There is some overlap in the categories listed in **Table 1**, since different interventions could be envisaged for the same land, and the integrated measures discussed by Smith et al. [8] notably result in increased carbon storage in soils. The category

Improved cropland management	1.4–2.3*
Increasing soil organic matter stocks	1.3–5.1*
Reduced deforestation and forest degradation	0.4–5.8*
Reduced conversion of grasslands	0.7*
Agroforestry	0.11–5.68**
Reduced conversion of coastal wetlands	0.11–2.25**
Biochar	0.03–6.60**
Cropland nutrient management N_2O	0.03–0.71**
Manure management N_2O and CH_4	0.01–0.26**
Improved rice cultivation CH_4	0.08–0.87**
Reduced enteric fermentation CH_4 (ruminants)	0.12–1.18**
Soil carbon sequestration in croplands	0.25–6.78**

*[8].
**[36].

Table 1.
Yearly global climate mitigation potential of different interventions (IPCC estimates in $GtCO_2eq$ yr).

'improved cropland management' includes practices such as reduced tillage, cover crops, perennials, water management and nutrient management.

4.1 Uncertainties in estimates and critical issues

The type of management system that farmers adopt, will substantially determine the capacity of soil to act as a carbon sink, and the extent to which agricultural land will contribute to GHG emissions. However, estimates regarding the potential of agricultural soil management practices to mitigate climate change vary considerably, and have been calculated in various manners. While Minasny et al. [71] estimate that raising soil organic matter could offset 20–35% of total GHG emissions, Schlesinger and Amundson [72] believe that the combined use of biochar and enhanced silicate weathering on agricultural land will not offset more than 5% of emissions. Differences in what is included in calculations, as well as in assumptions regarding anticipated conditions and future projections naturally affect conclusions. Biochar has attracted considerable interest for its ability to improve soil fertility and immobilize pollutants, while offering potential for long term storage of carbon [51]. However, the stability of biochar and its long-term impacts will ultimately depend on conditions that affect biochar aging [73]. With respect to upscaling enhanced silicate weathering as a climate mitigation strategy, uncertainties and possible negative environmental impacts need to be taken into account [74, 75].

Types of organic fertilizer that contain organic matter will directly contribute to soil organic carbon (SOC) content, but fungi and microbes contained in certain types of organic fertilizer, as well as impacts of pH and the proportions of other nutrients and micro-nutrients, will all affect the dynamics of soil biota and ecosystems. This leads to indirect positive or negative affects not only on fertility, water retention and resilience, but also on net GHG emissions (see e.g., Galic et al. [7], Walling et al. [47], Xu et al. [52]). Among other factors, annual precipitation significantly affects SOC dynamics [37, 76], and must be considered in arid and semi-arid regions.

4.2 Carbon sequestration

Carbon stocks in agricultural soils have been depleted worldwide, affecting productivity (see Droste et al. [77]). However, these losses do not all necessarily correspond to release of CO_2 into the atmosphere, and Chenu [78] therefore makes the distinction between carbon sequestration, which aims to counteract global warming, and carbon storage in soils. Numerous approaches are developed to enhance carbon sequestration. In New Zealand, for instance, 'flipping' is used for podzolized sandy soils with pasture grassland, to avoid water logging. Burying topsoil led to long term SOC preservation, while new organic matter could accumulate in the surface soil under these conditions [79]. However, as for most practices, impacts will be dependent on local circumstances, since disrupting soil ecosystems will alter SOC dynamics, thereby carbon contained in above-ground vegetation or root systems, while exposure of topsoil can lead to erosion. Madigan et al. [80] compare different approaches to managing pasture and argue that full-inversion tillage (FIT) during pasture renewal has potential in an Irish context, particularly when combined with re-seeding.

While many of the approaches aiming at carbon sequestration and reduction of GHG emissions [65] bring benefits for agriculture through soil improvement, increasing water retention, reducing agricultural runoff and effects of heat stress, as well as conserving ecosystems, there are nevertheless risks associated with the need to rapidly offset GHG emissions produced by the burning of fossil fuels. From the point of view of agricultural production, organic matter is urgently needed to counter loss of topsoil and soil degradation, while equally urgent ambitions to rapidly achieve long term sequestration of carbon at a large scale, will reduce the amount of organic material available. Some approaches to carbon sequestration keep soil organic matter (SOM) in soil layers and forms that remain available to vegetation, while others such as flipping [79] bury the SOM in lower layers in order to slow down metabolic processes. However, yet others aim to bind carbon in forms that are not bioavailable or bury it in deep sediment or geological layers that remove both carbon and nutrients contained in organic waste from biological cycles.

Soil microbial activity is beneficial to crops and supports agricultural productivity but can also result in a net increase of GHG emissions, depending on balance and conditions. The use of agricultural lime to improve acidic soils can either lead to increased release of CO_2 in the atmosphere, or to carbon sequestration. For instance, Bramble, Gouveia and Ramnarine [81] found that combining the application of agricultural lime with poultry litter prevented CO_2 emissions. Finally, it is important to also consider energy conservation in climate mitigation strategies. Soil organic content substantially affects energy requirements, and Hercher-Pasteur et al. [82] therefore argue that this should be included when calculating optimal uses for biomass.

5. Sustainable agricultural practices

Choice of fertilizer cannot be understood in isolation, but as part of overall soil and land management practices in agriculture. In the following, some examples of sustainable practices are given, that are supported by the use of organic fertilizers, but which can also enhance their benefits. Combinations of approaches lead to synergies, not only with respect to bioavailability of nutrients, but also with respect to water balance, prevention of erosion [37], pest and pathogen control, and resilience to other stressors. For instance, improving tillage practices and incorporating residue was found to increase water-use efficiency by 30%, rice–wheat yields by 5–37%, income by 28–40%, while reducing and GHG emissions by 16–25% [8]. Further options of interest include perennial crops [83–85], polyculture [86], mosaic landscapes [87] and the use of pollinator strips or other habitat [88, 89], which support crop productivity through ecological intensification [90].

5.1 The role of soil health and microbial activity

Loss of soil health exposes crops to various diseases [54]. Among the numerous challenges for soil health in arid and semi-arid regions is the risk of salinization [37, 54, 91], which is driven not only by evaporation and low precipitation, but also by use of synthetic fertilizers and reduced moisture retention in soils with low content of organic materials. Soil organisms are essential for soil fertility, by making nutrients available to crops. A healthy soil ecosystem decomposes organic matter, makes nutrients available, prevents nutrient leaching and fixes nitrogen. It also protects plants from pathogens [54], improves soil structure and promotes well-functioning root systems. However, microbial activity can contribute to GHG emissions, and net effects under different conditions therefore need to be carefully considered.

The fungal to bacteria biomass ration (F/B) is one of the important indicators of soil health. Optimal F/B ratios depend on intended use. While grains and vegetables require bacterial dominance or a balance between fungi and bacteria, orchard trees need a dominance of fungi, which are more effective at immobilizing nutrients, preventing leaching. For grasslands, higher F/B ratios are an indication of more sustainable systems, with less environmental impacts [92]. It should be noted that biomass in itself is not a complete indicator for fungal and microbial activity [92] and that the distribution across various depths is also important for fertility and GHG flux dynamics.

Fiodor et al. [54] point to the potential use of specific plant growth promoting microbes (PGPM) that protect against a wide range of stressors and pathogens, and which can be applied by methods such as inoculation. Although microbial communities can in many respects be interchangeable from a functional point of view, unique strains of PGPM that mitigate effects of biotic and abiotic stressors are especially relevant in the light of rapid climate change. Research on how organic fertilizers can support such microbes is therefore called for, as is research soil ecosystems and plant-microbial symbiotic relationships (see e.g., Porter and Sachs [93]). Impacts of antibiotic residues in organic fertilizer [10] also require attention.

5.2 Conservation agriculture

Soil conservation practices are needed for sustainable productivity [94]. Conservation agriculture (CA) conserves soil moisture and reduces both erosion and runoff, improving water quality, as well as promoting biodiversity and above-ground ecosystems [95], with potentials for pest control and pollination. CA has been found to reduce water use substantially, as well as decreasing energy inputs [96]. It is of particular interest under extreme climatic conditions, due to its ability to mitigate heat and water stress, thereby increasing crop yields [96] and resilience.

In an Indian context, Battacharya et al. [94] compared performance of CA practices with farms applying conventional tillage over a nine-year period, using a wide range of measurements for soil health and sustainability. In this Indian study, conservation agriculture was shown to increase SOC, while requiring low input. However, Palm et al. [95] underline that CA will not necessarily increase soil carbon sequestration in all contexts. In studies they reviewed, only about half reported increased sequestration with no-till practices. Furthermore, in Sub Saharan Africa, Palm et al. [95] found that lack of residues was a significant obstacle to implementing CA for smallholder farmers. Use of organic residues for soil amendment in these contexts competed with other uses that had higher values, primarily as fodder for livestock. They conclude that it is important to distinguish between high-input CA systems applied in large-scale mechanised farms, and which require large inputs of herbicides to control weeds, with conditions for smallholder systems in the tropics and subtropics.

5.3 Tillage practices

No-tillage systems and suitable cover crop management can improve SOC, total N, available P, exchangeable K-Mg, CEC, bulk density, soil penetration resistance, and substrate-induced respiration, as exemplified in a Japanese study concerning Andosols [97]. Inversely, tillage will increase microbial activity that contributes to emissions, accelerate decomposition, but the disturbance will reduce microbial communities over time [97]. However, according to the review made by Palm et al. [95], no-till systems in cooler and wetter climates are more likely to result in lower soil carbon and reduced crop yields.

5.4 Cover crops

Cover crops conserve water, moderate soil temperature, and help to control weeds. Cover crops can further increase fungal biomass and improve the biological structure of soil [92]. Long-term use of cover crops improves soil fertility through the accumulation of SOM [92]. Disrupting soils through tillage kills fungi, and therefore shifts the balance towards bacteria. Legume intercrops or cover crops can lead to higher soil carbon storage and slower decomposition in no-till rotation systems [95]. Palm et al. [95] found that while quality of organic inputs affected short-term carbon dynamics, it did not appear to substantially affect long-term storage. Quality could be modified by addition of lignin. Materials with a high carbon to N ratio could result in reduced crop yields, while residues with a lower C:N ratio, as in the case of legume residues and legume cover crops, increased N availability. Legumes are not only of interest for their N-fixing properties, but for other facilitation effects as well [98–100].

5.5 Agroforestry

Agroforestry brings benefits for soil fauna and generally improves soil quality [101–103], and soil organic carbon sequestration [51, 104]. Depending on conditions, reduced light can affect yields of crops that are grown with trees, but agroforestry is also deliberately used to provide shade and create beneficial microclimates to mitigate heat stress and loss of water through evapotranspiration, as well as to adjust for lower or more variable rainfall [105], which is highly relevant for arid and semi-arid regions. With global warming, weather systems will contain more energy, and agroforestry therefore can play a role in preventing erosion and loss of soil from wind [37], as well as from extreme rainfall. Agroforestry systems can offer valuable habitat for pollinators and fauna essential for pest control, but trees should be selected for climate resilience and the precise combinations of species of orchards, crops or other vegetation in these systems needs to be considered, as well as spacing, orientation and adjustment to topography.

6. Water conservation and pollution prevention

Major landscape changes, with loss and deterioration of wetlands [26, 106], mean that nutrient flows from agriculture rapidly move on into the oceans, destabilizing ecosystems [107]. Drainage, to claim land for agriculture or other purposes, and extensive irrigation in agriculture cause wetlands to dry [108], while other drivers of wetland loss are urbanisation and surface sealing for road networks, industrial use of water and large dams. With climate change, water is no longer released gradually over the year through snow smelting, and forest fires [41], use of woodlands for fuel or commercial logging create additional disruptions in the water systems on which wetlands depend [109]. The amount of carbon stored in wetlands and peatlands constitutes in the order of 30–40% of terrestrial carbon [110, 111].

According to UN Water, 72% of all water withdrawals globally are used by agriculture [112]. Besides practices such as no-till, reduced till, cover crops or terracing and contour farming to retain water and reduce erosion [37], leaving crop residue on the surface also serves these purposes [113]. Importantly, demand for water can be further reduced by supporting complex agricultural landscapes that include trees and other vegetation, and by shifting to crops and cultivars that require less water. Alongside conventional approaches to water conservation such as drip irrigation, such approaches are necessary to address the water crisis, which will in many regions be aggravated by climate change [39–41]. For arid and semi-arid regions in particular, conservation agriculture and other sustainable practices are crucial for their role in preserving soil moisture and reducing irrigation needs. Both organic fertilizers and other methods of increasing SOM play a role in reclaiming land and combatting desertification [8, 37, 59, 99, 114–116]. Several solutions to the issue of polluted water [26, 106] have been suggested, including phytoremediation or the use of agricultural waste to serve as biosorbants [117–119].

Bhattacharyya et al. [120] suggest nutrient budgeting as an effective approach to preventing soil-water-air contamination from crop-livestock systems. Excess nutrients do not only impact rivers, lakes and coastal waters, but also affect groundwater quality [28, 29]. Nutrient surpluses are linked to use of fertilizers and manure, as well as to low nutrient utilization efficiency of plants. Leaching, runoff and erosion are therefore all significant for sustainable agricultural practices. In this respect, a slower release of nutrients and improvements in soil structure are important potential benefits of organic fertilizers, compared to chemical fertilizers. Contributions to soil and ecosystem health of sustainable practices reduce the need for pesticides to control pests and pathogens, thereby increasing availability of good quality water [49] and protecting the world's oceans [121, 122].

The various interlinkages and trade-offs that need to be considered in use of water resources are acknowledged in European policy on the water, energy, food, and ecosystems (WEFE) nexus [123], as well as in recent research in this field [124–127]. Both general conflicts in demands concerning use of land and resources, and water scarcity, in particular, affect the arid and semi-arid regions of the MENA region. For these regions, land management must pay greater attention to how soil health and quality affects water retention. Degraded soils have poor water retention capacity, demand more fertilizer, and are less able to contribute to carbon sequestration. A more holistic view of land and soil management can also mitigate effects of stress caused by heat, extreme weather events and increased climate variability.

7. Transition issues

Conservation agriculture can lead to yield benefits, but improvements may not be noticeable in the initial years [94]. In a Swedish context, examining various sites over a period of 54 years, Droste et al. [77] find that increasing SOC leads to long-term yield stability and resilience, which is important in view of accelerating climate change. However, adopting sustainable management practices can come at the cost of short-term productivity. Policy changes to support the transition are therefore recommended [77, 128]. To minimise initial economic impacts for farmers of conversion, Yigezu et al. [46] and Tu et al. [129] recommend transition strategies that involve gradually reducing conventional inputs.

Sustainable agricultural practices achieve control of pests and pathogens without damaging the environment, but these practices are also largely dependent on healthy soil biota and rich ecosystems in the agricultural landscape. Agricultural soils have been affected by numerous sources of pollution [130]. Soil management practices and use of chemicals will have negative effects on many soil invertebrates and microbes [131, 132] but will favour others. The net effect is therefore not only loss of important strains of soil biota or total mass, but the creation of imbalances in microbial communities that can have detrimental effects for plant health and crop yields.

Since soil health and ecosystems have been damaged by prior unsustainable practices, including use of synthetic fertilizers and pesticides, restoring health takes time, and processes of remediation and restoration are therefore crucial [59, 77, 132–134]. The ability of new cultivars to benefit from plant-microbial symbiosis has been affected by selection of cultivars for other traits, and by reduced dependence on this symbiosis through the use of synthetic fertilizers [93]. Transition to sustainable farming with organic fertilizers should therefore also consider the choice of suitable cultivars and heritage varieties that retain the ability to fully benefit from improved soil health.

8. Smallholder farming and sustainable agriculture

It is difficult to evaluate the magnitude of smallholder and subsistence farming world-wide, since it is frequently undertaken in regions with limited statistics, on fragmented or mixed-use plots where land-use can be difficult to identify from satellite images. In many contexts, it is not necessarily the primary occupation of the farmer. Despite its marginal position in debates on agricultural productivity, smallholder farming plays a vital role for biodiversity, food security, human health, equity and climate resilience, since value is not lost in the distribution chain but stays with producers and their communities. Locally sourced food reduces community vulnerability to disruptions in the food supply chain, due to disasters, logistics failures, financial crises, or armed conflict. The latter consideration is significant for the MENA region, where several countries are affected by conflict or volatility [33]. Food systems worldwide are exposed to numerous disruptions, which will increase as a result of climate change and environmental degradation [69]. Smallholder farmers are particularly vulnerable to such shocks and have difficulties making adequate choices in the face of uncertainty [21, 22, 31]. To address such challenges, Kim et al. [70] suggest a land-water-nutrient nexus (LWNN) approach (see also Jat et al. [96] for strategies from an Indian context). Crop diversification can be a strategy to meet the double uncertainty of price fluctuations and crop failures [135], and polycultures also have environmental benefits. However, food processing industries and international markets tend to be oriented towards monocultures, and smallholder farmers can be obligated by contracts to produce particular crops.

Low-input smallholder production systems are one of the dominant food production systems globally [136]. In an Ethiopian case, Baudron et al. [136] observe that complex agricultural landscapes that incorporate trees offer better overall livelihoods for farmers, lead to better carbon balances, as well as being more resilient both to fluctuation in input prices and to climate stress. They further underline that low-input farming with resource-saving practices can increase profitability for farmers more than yield optimization, while yield stability is another important consideration for smallholders.

Baudron et al. [136] therefore argue for an increased attention to agricultural practices that support synergies between agriculture and biodiversity, rather than presenting the situation as an irreducible choice between 'land sparing'—aiming to reduce demand for land through intensification— and 'land sharing', assuming loss in yields, as a consequence of practices that are more favourable to wildlife and biodiversity. Baudron et al. emphasize the reliance of low-input smallholder agricultural production on ecosystem services provided by biodiverse ecosystems, and further point to the crucial role of ecosystem services to maintain soil fertility, pollination, and for pest and disease control [136, 137].

Despite the benefits that low-input farming can bring [138], barriers include lack of locally relevant expertise, and the time needed to rehabilitate soils degraded by use of synthetic fertilizers and pesticides. Subsidy systems may support heavily mechanised and chemical-intensive agriculture [3, 14], with questionable benefits for smallholder farmers. Further barriers in transitioning to sustainable agricultural practices are access to markets, and global food systems structured to favour monoculture of particular crops and cultivars.

9. Conclusions

In view of the numerous factors that influence outcomes for the use of organic fertilizers, locally tailored strategies that combine approaches to enhance soil health and sustainable land management would be recommended. However, sufficiently detailed data is still lacking on how different management practices affect yields and environmental impacts depending on local conditions, particularly in the global South. Citizen science has the potential to offer a better evidence base for farmers' choices, but the structure of many citizen science projects rarely supports longer term collaboration and dialogue with smallholder farmers in the global South [139, 140]. In addition, smallholder farmers may not be able to afford individualised consulting, and agronomists may lack expertise applicable to low-input agriculture. Transitioning to sustainable practices is knowledge intensive [44], and this is therefore an area where international networking with academic institutions could play a significant role in supporting climate adaptation and mitigation efforts. Exchange of knowledge among farmers [141] and farmers' organizations can also play a role for mobilizing resources and expertise, but such potential contributions will depend on the orientation of the organization [142].

Among other implications of the current climate crisis, a narrow focus on crop yields is not sufficient, since outcomes of fertilizer application are usually estimated under optimal or normal growing conditions. Increased weather variability and the ensuing risk of crop failure, means that greater attention must be devoted to resilience, and the capacity to cultivate under unpredictable and less than optimal conditions. This in turn means, for instance, that effects on root growth, the capacity of root systems to absorb water and nutrients under extreme conditions, as well as the capacity of the soil to retain water and nutrients over longer periods of time all become critical factors. Also, rather than considering fertilizer application merely from the view of inputs and short-term yields, and besides measures such as C:N ratios, we need to take on a more holistic view, looking at how choice of fertilizer relates to nutrient absorption efficiency, drought resistance of root systems [143], soil health, land degradation, water management and ecological intensification. Future shortages of P [144–146], loss of arable land [37], decline in soil carbon [147], as well as widespread decline in soil fertility driven by industrial practices in agriculture, point to the important role of organic fertilizers. However, availability of organic material is constrained by competing demands on biomass and land for industrial and carbon sequestration purposes, while contamination of organic waste and wastewater [10, 118, 148] poses an issue for possible circular approaches. To generalise the use of organic fertilizers, redesign of food systems and policy changes are therefore required, adopting a more comprehensive approach to the complex interlinkages that are involved.

Acknowledgements

The Swedish Research Council for Sustainable Development, FORMAS (project number 2017-01375) has contributed to APC for this publication.

Author details

Helen Avery
Centre for Environmental and Climate Science/Centre for Advanced Middle Eastern Studies, Lund University, Lund, Sweden

*Address all correspondence to: helen.avery@cme.lu.se

References

[1] Sradnick A, Feller C. A typological concept to predict the nitrogen release from organic fertilizers in farming systems. Agronomy. 2020;**10**(9):1448

[2] Rayne N, Aula L. Livestock manure and the impacts on soil health: A review. Soil Systems. 2020;**4**(4):64. DOI: 103390/soilsystems4040064

[3] Jain M, Solomon D, Capnerhurst H, Arnold A, Elliott A, Kinzer AT, et al. How much can sustainable intensification increase yields across South Asia? A systematic review of the evidence. Environmental Research Letters. 2020;**15**(8):083004

[4] Mengqi Z, Shi A, Ajmal M, Ye L, Awais M. Comprehensive review on agricultural waste utilization and high-temperature fermentation and composting. Biomass Conversion and Biorefinery. 2021:1-24. DOI: 101007/s13399-021-01438-5. The article is available online at: https://link.springer.com/article/10.1007/s13399-021-01438-5

[5] Shaji H, Chandran V, Mathew L. Organic fertilizers as a route to controlled release of nutrients. In: Lewu FB, Volova T, Thomas S, Rakhimol KR, editors. Controlled Release Fertilizers for Sustainable Agriculture. Cambridge, Massachusetts: Academic Press; 2021. pp. 231-245. DOI: 101016/B978-0-12-819555-000013-3. ISBN 9780128195550. https://www.sciencedirect.com/book/9780128195550/controlled-release-fertilizers-for-sustainable-agriculture#book-info

[6] Shi W, Zhao HY, Chen Y, Wang JS, Han B, Li CP, et al. Organic manure rather than phosphorus fertilization primarily determined asymbiotic nitrogen fixation rate and the stability of diazotrophic community in an upland red soil. Agriculture, Ecosystems Environment. 2021;**319**:107535

[7] Galic M, Mesic M, Zgorelec Z. Influence of organic and mineral fertilization on soil greenhouse gas emissions: A review. Agriculturae Conspectus Scientificus. 2020;**85**(1):1-8

[8] Smith P, Nkem J, Calvin K, Campbell D, Cherubini F, Grassi G, et al. Interlinkages between desertification, land degradation, food security and greenhouse gas fluxes: Synergies, trade-offs and integrated response options In: Shukla PR, Skea J, Calvo Buendia E, Masson-Delmotte V, Portner H-O, Roberts DC, editors. Climate Change and Land: An IPCC special report on climate change, desertification, land degradation, sustainable land management, food security, and greenhouse gas fluxes in terrestrial ecosystems. New York: Intergovernmental Panel on Climate Change 2019

[9] Romanos DM, Nemer N, Khairallah Y, Abi Saab MT. Application of sewage sludge for cereal production in a Mediterranean environment (Lebanon). International Journal of Recycling Organic Waste in Agriculture. 2021;**10**(3):233-244. DOI: 1030486/IJROWA202119037391098

[10] Mahjoory Y, Aliasgharzad N, Moghaddam G, Bybordi A. Long-term application of manure alters culturable soil microbial populations and leads to occurrence of antibiotic resistant bacteria. Soil and Sediment Contamination: An International Journal. 2021:1-15. DOI: 101080/1532038320211961122. The article is available online at: https://www.tandfonline.com/doi/full/10.1080/15320383.2021.1961122

[11] Li S, Li J, Li G, Li Y, Yuan J, Li D. Effect of different organic fertilizers application on soil organic matter properties. Compost Science Utilization. 2017;**25**(suppl. 1):S31-S36

[12] Soleymani F, Ahmadvand G, Safari Sinegani AA. Effect of chemical, biological and organic fertilizers on growth indices, yield and yield components of sunflower under optimum and deficit irrigation. Journal of Agricultural Science and Sustainable Production. 2017;**27**(2):19-35

[13] Parmar DK, Sharma A, Chaddha S, Sharma V, Vermani A, Mishra A, et al. Increasing potato productivity and profitability through integrated plant nutrient system in the north-western Himalayas. Potato Journal. 2007;**34**(3-4):209-215

[14] Rashid S, Dorosh PA, Malek M, Lemma S. Modern input promotion in sub-Saharan Africa: Insights from Asian green revolution. Agricultural Economics. 2013;**44**(6):705-721

[15] Nelson ARLE, Ravichandran K, Antony U. The impact of the green revolution on indigenous crops of India. Journal of Ethnic Foods. 2019;**6**(1):1-10

[16] Hou D, Bolan NS, Tsang DC, Kirkham MB, O'Connor D. Sustainable soil use and management: An interdisciplinary and systematic approach. Science of the Total Environment. 2020;**729**:138961

[17] Giltrap D, Cavanagh J, Stevenson B, Ausseil AG. The role of soils in the regulation of air quality. Philosophical Transactions of the Royal Society B. 2021;**376**(1834):20200172

[18] Lohmann U, Friebel F, Kanji ZA, Mahrt F, Mensah AA, Neubauer D. Future warming exacerbated by aged-soot effect on cloud formation. Nature Geoscience. 2020;**13**(10):674-680

[19] Boloorani AD, Najafi MS, Mirzaie S. Role of land surface parameter change in dust emission and impacts of dust on climate in Southwest Asia. Natural Hazards. 2021;**109**:111-132. DOI: 10.1007/s11069-021-04828-0

[20] IPCC. Summary for policymakers. In: Masson-Delmotte V, Zhai P, Pirani A, Connors SL, Péan C, Berger S, editors. Climate Change 2021: The Physical Science Basis Contribution of Working Group I to the Sixth Assessment Report of the Intergovernmental Panel on Climate Change. 2021. ISBN: 978-92-9169-158-6

[21] Menghistu HT, Abraha AZ, Tesfay G, Mawcha GT. Determinant factors of climate change adaptation by pastoral/agro-pastoral communities and smallholder farmers in sub-Saharan Africa: A systematic review. International Journal of Climate Change Strategies and Management. 2020;**12**(3): 305-321. DOI: 101108/IJCCSM-07-2019-0049

[22] Waldman KB, Guido Z, Todd PM, Evans TP, Carrico A, Attari SZ. Reorienting climate decision making research for smallholder farming systems through decision science. Current Opinion in Environmental Sustainability. 2021;**52**:92-99

[23] Woertz E. Food and water security in West Asia. In: Allan T, Bromwich B, Keulertz M, Colman A, editors. Oxford, New York: The Oxford Handbook of Food, Water and Society, Oxford University Press; 2018. DOI: 101093/oxfordhb/9780190669799.01329

[24] FAO. Aquastat. Available from: https://wwwfaoorg/aquastat/statistics [Accessed: October 14, 2021]

[25] Nouri H, Stokvis B, Borujeni SC, Galindo A, Brugnach M, Blatchford ML, et al. Reduce blue water scarcity and increase nutritional and economic water productivity through changing the cropping pattern in a catchment. Journal of Hydrology. 2020;**588**:125086. DOI: 101016/jjhydrol2020125086

[26] Salman NA, Al-Saas HT, Al-Imarah FJ. The status of pollution in the southern marshes of Iraq: A short

review. In: Jawad LA, editor. Southern Iraq's marshes Their Environment and Conservation. Coastal Research Library 36. Cham: Springer; 2021. pp. 505-516

[27] Al-Saad HT, Al-Imarah FJ. Pesticides in the waters, sediments, and biota of the Shatt Al-Arab River for the period 1980-2017. In: Jamad LA, editor. Tigris and Euphrates Rivers: Their Environment from Headwaters to Mouth. Cham: Springer; 2021. pp. 299-308

[28] Jalali M, Jalali M. An investigation on groundwater geochemistry changes after 17 years: A case study from the west of Iran. Environmental Earth Sciences. 2020;**79**(15):1-15. DOI: 101007/s12665-020-09114-z

[29] Almasri MN, Judeh TG, Shadeed SM. Identification of the nitrogen sources in the Eocene aquifer area (Palestine). Water. 2020;**12**(4):1121. DOI: 103390/w12041121

[30] UNDESA. Available from: Population Division | (unorg). [Accessed: October 14, 2021]

[31] Al Dirani A, Abebe GK, Bahn RA, Martiniello G, Bashour I. Exploring climate change adaptation practices and household food security in the Middle Eastern context: A case of small family farms in Central Bekaa, Lebanon. Food Security. 2021;**13**(4):1029-1047. DOI: 101007/s12571-021-01188-2

[32] Imai KS, Gaiha R, Garbero A. Poverty reduction during the rural–urban transformation: Rural development is still more important than urbanisation. Journal of Policy Modeling. 2017;**39**(6):963-982

[33] Crisis group. Available from: Crisis Group. [Accessed: October 14, 2021]

[34] UNHCR Global Appeal. 2021. Available from: https://reportingunhcrorg/sites/default/files/ga2021/pdf/Global_Appeal_2021_full_lowrespdf. [Accessed: October 14, 2021]

[35] Global Humanitarian Overview 2022. Geneva: UNOCHA (United Nations Office for the Coordination of Humanitarian Affairs). 2021. p. 304. Available online at https://www.unocha.org/sites/unocha/files/Global%20Humanitarian%20Overview%202022.pdf

[36] Shukla PR, Skea J, Slade R, van Diemen R, Haughey E, Malley J, et al. Technical Summary. In: Climate Change and Land: an IPCC special report on climate change, desertification, land degradation, sustainable land management, food security, and greenhouse gas fluxes in terrestrial ecosystems [P.R. Shukla, J. Skea, E. Calvo Buendia, V. Masson-Delmotte, H.-O. Pörtner, D. C. Roberts, P. Zhai, R. Slade, S. Connors, R. van Diemen, M. Ferrat, E. Haughey, S. Luz, S. Neogi, M. Pathak, J. Petzold, J. Portugal Pereira, P. Vyas, E. Huntley, K. Kissick, M. Belkacemi, J. Malley, (eds.)]. 2019. In press.

[37] IPCC Special Report on Climate Change and Land. Chapter 3. Available from: Chapter 3 : Desertification — Special Report on Climate Change and Land (ipccch). [Accessed: October 14, 2021]

[38] Zittis G, Hadjinicolaou P, Almazroui M, Bucchignani E, Driouech F, El Rhaz K, et al. Business-as-usual will lead to super and ultra-extreme heatwaves in the Middle East and North Africa. npj Climate and Atmospheric Science. 2021;**4**(1):1-9

[39] Zittis G, Bruggeman A, Lelieveld J. Revisiting future extreme precipitation trends in the Mediterranean. Weather and Climate Extremes. 2021;**34**:100380. DOI: 01016/jwace2021100380

[40] Ajjur SB, Al-Ghamdi SG. Evapotranspiration and water availability response to climate change in the Middle East and North Africa. Climatic Change. 2021;**166**(3):1-18

[41] Tomaszkiewicz MA. Future seasonal drought conditions over the CORDEX-MENA/Arab domain. Atmosphere. 2021;**12**(7):856

[42] Cooper R. Climate Change Risks and Opportunities in the Middle East and North AfricaK4D Helpdesk Report. Brighton, UK: Institute of Development Studies; 2020

[43] Khouri N, Breisinger C, Eldidi H. Can MENA reach the sustainable development goals? An overview of opportunities and challenges for food and nutrition security. In: Mergos G, Papanastassiou M, editors. Food Security and Sustainability. Cham: Palgrave Macmillan. 2017:175-191. DOI: 10.1007/978-3-319-40790-6_10

[44] Koç G, Uzmay A. Determinants of dairy farmers' likelihood of climate change adaptation in the Thrace Region of Turkey. Environment, Development and Sustainability. 2021:1-22. https://link.springer.com/article/10.1007/s10668-021-01850-x

[45] Delfiyan F, Yazdanpanah M, Forouzani M, Yaghoubi J. Farmers' adaptation to drought risk through farm–level decisions: The case of farmers in Dehloran county, Southwest of Iran. Climate and Development. 2021;**13**(2):152-163

[46] Yigezu YA, El-Shater T, Boughlala M, Devkota M, Mrabet R, Moussadek R. Can an incremental approach be a better option in the dissemination of conservation agriculture? Some socioeconomic justifications from the drylands of Morocco. Soil and Tillage Research. 2021;**212**:105067

[47] Walling E, Vaneeckhaute C. Greenhouse gas emissions from inorganic and organic fertilizer production and use: A review of emission factors and their variability. Journal of Environmental Management. 2020;**276**:111211

[48] Grizzetti B, Billen G, Davidson EA, Winiwarter W, de Vries W, Fowler D, et al. Global nitrogen and phosphorus pollution. In: Sutton MA et al., editors. Just Enough Nitrogen. Cham: Springer; 2020. pp. 421-431. DOI: 101007/978-3-030-58065-0_28

[49] Parizad S, Bera S. The effect of organic farming on water reusability sustainable ecosystem, and food toxicity. Environmental Science and Pollution Research. 2021:1-12. DOI: 101007/s11356-021-15258-7

[50] Pörtner HO, Scholes RJ, Agard J, Archer E, Arneth A, Bai X, et al. Scientific Outcome of the IPBES-IPCC Co-sponsored Workshop on Biodiversity and Climate Change. Bonn, Germany: IPBES Secretariat; 2021. DOI: 105281/zenodo4659158

[51] Dooley K, Harrould-Kolieb E, Talberg A. Carbon-dioxide removal and biodiversity: A threat identification framework. Global Policy. 2021;**12**:34-44

[52] Xu X, Xia Z, Liu Y, Liu E, Müller K, Wang H, et al. Interactions between methanotrophs and ammonia oxidizers modulate the response of in situ methane emissions to simulated climate change and its legacy in an acidic soil. Science of the Total Environment. 2021;**752**:142225. DOI: 101016/jscitotenv2020142225

[53] Arneth A, Denton F, Agus F, Elbehri A, Erb K, Osman Elasha B, et al. Arneth A, Denton F, Agus F, Elbehri, A, Erb K, Osman Elasha B, Rahimi M, Rounsevell M, Spence A, Valentini R Framing and Context. In: Climate Change and Land: An IPCC Special Report on Climate Change, Desertification, Land Degradation, Sustainable Land Management, Food Security, and Greenhouse Gas Fluxes in Terrestrial Ecosystems. Shukla PR, Skea J, Buendia EC, Masson-Delmotte V, Pörtner H-O, Roberts DC. editors. Geneva, Switzerland: The Intergovernmental

Panel on Climate Change (IPCC); 2019. p. 77-12

[54] Fiodor A, Singh S. The contrivance of plant growth promoting microbes to mitigate climate change impact in agriculture. Microorganisms. 2021;**9**(9): 1841. DOI: 103390/microorganisms 9091841

[55] IPCC. Summary for policymakers. In: Climate Change and Land: An IPCC Special Report on Climate Change, Desertification, Land Degradation, Sustainable Land Management, Food Security, and Greenhouse Gas Fluxes in Terrestrial Ecosystems. Shukla PR, Skea J, Buendia EC, Masson-Delmotte V, Pörtner H-O, Roberts DC, et al. editors. Geneva, Switzerland: The Intergovernmental Panel on Climate Change (IPCC); 2019

[56] Lenton TM, Rockström J, Gaffney O, Rahmstorf S, Richardson K, Steffen W, et al. Climate tipping points—Too risky to bet against. Nature. 2019;**575**:592-595

[57] Galloway JN, Bleeker A, Erisman JW. The human creation and use of reactive nitrogen: A global and regional perspective. Annual Review of Environment and Resources. 2021;**46**(18):1-34. DOI: 101146/annurev-environ-012420-045120

[58] Tian H, Yang J, Xu R, Lu C, Canadell JG, Davidson EA, et al. Global soil nitrous oxide emissions since the preindustrial era estimated by an ensemble of terrestrial biosphere models: Magnitude, attribution, and uncertainty. Global Change Biology. 2019;**25**:640-659

[59] Arneth A, Olsson L, Cowie A, Erb KH, Hurlbert M, Kurz WA, et al. Restoring degraded lands. Annual Review of Environment and Resources. 2021;**46**(20):1-31. DOI: 101146/annurev-environ-012320-054809

[60] Humphreys J, Brye KR, Rector C, Gbur EE. Methane emissions from rice across a soil organic matter gradient in Alfisols of Arkansas, USA. Geoderma Regional. 2019;**16**:e00200

[61] da Silva CA, Quintana BG, Janusckiewicz ER, de Figueiredo BL, da Silva ME, Reis RA, et al. How do methane rates vary with soil moisture and compaction, N compound and rate, and dung addition in a tropical soil? International Journal of Biometeorology. 2019;**63**(11):1533-1540. DOI: 10.1007/s00484-018-1641-0

[62] Lade SJ, Steffen W, De Vries W, Carpenter SR, Donges JF, Gerten D, et al. Human impacts on planetary boundaries amplified by Earth system interactions. Nature Sustainability. 2020;**3**(2):119-128

[63] Wu Q, Hao J, Yu Y, Liu J, Li P, Shi Z, et al. The way forward confronting eco-environmental challenges during land-use practices: A bibliometric analysis. Environmental Science and Pollution Research. 2018;**25**(28): 28296-28311

[64] Aznar-Sánchez JA, Piquer-Rodríguez M, Velasco-Muñoz JF, Manzano-Agugliaro F. Worldwide research trends on sustainable land use in agriculture. Land Use Policy. 2019;**87**:104069

[65] Sykes AJ, Macleod M, Eory V, Rees RM, Payen F, Myrgiotis V, et al. Characterising the biophysical, economic and social impacts of soil carbon sequestration as a greenhouse gas removal technology. Global Change Biology. 2020;**26**(3):1085-1108

[66] Hong C, Burney JA, Pongratz J, Nabel JE, Muelle, ND, Jackson RB,

Davis SJ. Global and regional drivers of land-use emissions in 1961-2017. Nature. 2021;**589**(7843):554-561

[67] Lv T, Wang L, Xie H, Zhang X, Zhang Y. Exploring the global research trends of land use planning based on a bibliometric analysis: Current status and future prospects. Land. 2021;**10**(3):304

[68] Nalau J, Verrall B. Mapping the evolution and current trends in climate change adaptation science. Climate Risk Management. 2021;**32**:100290

[69] Béné C. Resilience of local food systems and links to food security–A review of some important concepts in the context of COVID-19 and other shocks. Food Security. 2020;**12**:805-822. DOI: 10.1007/s12571-020-01076-1. https://link.springer.com/article/10.1007/s12571-020-01076-1

[70] Kim DG, Grieco E, Bombelli A, Hickman JE, Sanz-Cobena A. Challenges and opportunities for enhancing food security and greenhouse gas mitigation in smallholder farming in sub-Saharan Africa: A review. Food Security. 2021;**13**:457-476. DOI: 101007/s12571-021-01149-9

[71] Minasny B, Malone BP, McBratney AB, Angers DA, Arrouays D, Chambers A, et al. Soil carbon 4 per mille. Geoderma. 2017;**292**:59-86. DOI: 101016/jgeoderma201701002

[72] Schlesinger WH, Amundson R. Managing for soil carbon sequestration: Let's get realistic. Global Change Biology. 2019;**25**(2):386-389

[73] Wang L, O'Connor D, Rinklebe J, Ok YS, Tsang DC, Shen Z, et al. Biochar aging: Mechanisms, physicochemical changes, assessment, and implications for field applications. Environmental Science Technology. 2020;**54**(23): 14797-14814

[74] Edwards DP, Lim F, James RH, Pearce CR, Scholes J, Freckleton RP, et al. Climate change mitigation: Potential benefits and pitfalls of enhanced rock weathering in tropical agriculture. Biology Letters. 2017;**13**(4):20160715

[75] Andrews MG, Taylor LL. Combating climate change through enhanced weathering of agricultural soils. Elements. 2019;**15**(4):253-258

[76] Luo Z, Feng W, Luo Y, Baldock J, Wang E. Soil organic carbon dynamics jointly controlled by climate, carbon inputs, soil properties and soil carbon fractions. Global Change Biology. 2017;**23**(10):4430-4439. DOI: 101111/gcb13767

[77] Droste N, May W, Clough Y, Börjesson G, Brady M, Hedlund K. Soil carbon insures arable crop production against increasing adverse weather due to climate change. Environmental Research Letters. 2020;**15**(12):124034

[78] Chenu C, Angers DA, Barré P, Derrien D, Arrouays D, Balesdent J. Increasing organic stocks in agricultural soils: Knowledge gaps and potential innovations. Soil and Tillage Research. 2019;**188**:41-52

[79] Schiedung M, Tregurtha CS, Beare MH, Thomas SM, Don A. Deep soil flipping increases carbon stocks of New Zealand grasslands. Global Change Biology. 2019;**25**(7):2296-2309. DOI: 101111/gcb14588

[80] Madigan AP, Zimmermann J, Krol DJ, Williams M, Jones MB. Full Inversion Tillage (FIT) during pasture renewal as a potential management strategy for enhanced carbon sequestration and storage in Irish grassland soils. Science of the Total Environment. 2022;**805**:150342. DOI: 10.1016/j.scitotenv.2021.150342

[81] Bramble DSE, Gouveia GA, Ramnarine R. Organic residues and ammonium effects on CO_2 emissions and soil quality indicators in limed acid tropical soils. Soil Systems. 2019;**3**(1):16. DOI: 103390/soilsystems 3010016

[82] Hercher-Pasteur J, Loiseau E, Sinfort C, Hélias A. Energetic assessment of the agricultural production system: A review. Agronomy for Sustainable Development. 2020;**40**(4):1-23

[83] Tesdell O, Othman Y, Dowani Y, Khraishi S, Deeik M, Muaddi F, et al. Envisioning perennial agroecosystems in Palestine. Journal of Arid Environments. 2020;**175**:104085. DOI: 101016/jjaridenv2019104085

[84] Isgren E, Andersson E, Carton W. New perennial grains in African smallholder agriculture from a farming systems perspective: A review. Agronomy for Sustainable Development. 2020;**40**(1):1-14. DOI: 1007/s13593-020-0609-8

[85] Schlautman B, Bartel C, Diaz-Garcia L, Fei S, Flynn S, Haramoto E, et al. Perennial groundcovers: An emerging technology for soil conservation and the sustainable intensification of agriculture. Emerging Topics in Life Sciences. 2021;**5**(2): 337-347

[86] Adamczewska-Sowińska K, Sowiński J. Polyculture management: A crucial system for sustainable agriculture development. In: Meena R, editor. Soil Health Restoration and Management. Singapore: Springer; 2020. pp. 279-319. DOI: 101007/978-981-13-8570-4_8

[87] Aguilera E, Díaz-Gaona C, García-Laureano R, Reyes-Palomo C, Guzmán GI, Ortolani L, et al. Agroecology for adaptation to climate change and resource depletion in the Mediterranean region: A review. Agricultural Systems. 2020;**181**:102809. DOI: 101016/jagsy2020102809

[88] Christmann S, Aw-Hassan A, Güler Y, Sarisu HC, Bernard M, Smaili MC, et al. Two enabling factors for farmer-driven pollinator protection in low-and middle-income countries. International Journal of Agricultural Sustainability. 2021:1-14. DOI: 101080/14735903 20211916254

[89] Christmann S, Bencharki Y, Anougmar S, Rasmont P, Smaili MC, Tsivelikas A, et al. Farming with Alternative Pollinators benefits pollinators, natural enemies, and yields, and offers transformative change to agriculture. Scientific Reports. 2021;**11**(1):1-10. DOI: 101038/s41598-021-97695-5

[90] Garibaldi LA, Pérez-Méndez N, Garratt MP, Gemmill-Herren B, Miguez FE, Dicks LV. Policies for ecological intensification of crop production. Trends in Ecology Evolution. 2019, 2019;**34**(4):282-286

[91] Sharifi P, Shorafa M, Mohammadi MH. Investigating the effects of cow manure, vermicompost and Azolla fertilizers on hydraulic properties of saline-sodic soils. International Journal of Recycling Organic Waste in Agriculture. 2021;**10**(1):43-51. DOI: 1030486/IJROWA202018934321039

[92] Nakamoto T, Komatsuzaki M, Hirata T, Araki H. Effects of tillage and winter cover cropping on microbial substrate-induced respiration and soil aggregation in two Japanese fields. Soil Science and Plant Nutrition. 2012;**58**(1):70-82

[93] Porter SS, Sachs JL. Agriculture and the disruption of plant–microbial symbiosis. Trends in Ecology Evolution. 2020;**35**(5):426-439. DOI: 101016/jtree202001006

[94] Bhattacharya P, Maity PP, Mowrer J, Maity A, Ray M, Das S, et al. Assessment of soil health parameters and application of the sustainability index to fields under conservation agriculture for 3, 6, and 9 years in India. Heliyon. 2020;**6**(12):e05640

[95] Palm C, Blanco-Canqui H, DeClerck F, Gatere L, Grace P. Conservation agriculture and ecosystem services: An overview. Agriculture, Ecosystems Environment. 2014;**187**: 87-105. DOI: 101016/jagee201310010

[96] Jat HS, Datta A, Choudhary M, Sharma PC, Jat ML. Conservation Agriculture: Factors and drivers of adoption and scalable innovative practices in Indo-Gangetic plains of India—A review. International Journal of Agricultural Sustainability. 2021;**19**(1):40-55

[97] Wulanningtyas HS, Gong Y, Li P, Sakagami N, Nishiwaki J, Komatsuzaki M. A cover crop and no-tillage system for enhancing soil health by increasing soil organic matter in soybean cultivation. Soil and Tillage Research. 2021;**205**:104749

[98] Iannetta PPM, Hawes C, Begg GS, Maaß H, Ntatsi G, Savvas D, et al. A multifunctional solution for wicked problems: Value-chain wide facilitation of legumes cultivated at bioregional scales is necessary to address the climate-biodiversity-nutrition nexus. Frontiers in Sustainable Food Systems. 2021;**5**:692137. DOI: 103389/fsufs 2021692137

[99] Abdelfattah MA, Rady MM, Belal HEE, Belal EE, Al-Qthanin R, Al-Yasi HM, et al. Revitalizing fertility of nutrient-deficient virgin sandy soil using leguminous biocompost boosts Phaseolus vulgaris performance. Plants. 2021;**10**(8):1637. DOI: 103390/ plants10081637

[100] Salama HS, Nawar AI, Khalil HE, Shaalan AM. Improvement of maize productivity and N use efficiency in a no-tillage irrigated farming system: Effect of cropping sequence and fertilization management. Plants. 2021, 2021;**10**(7):1459. DOI: 103390/plants 10071459

[101] Cardinael R, Mao Z, Chenu C, Hinsinger P. Belowground functioning of agroforestry systems: Recent advances and perspectives. Plant Soil. 2020;**453**: 1-13. DOI: 10.1007/s11104-020-04633-x

[102] Marsden C, Martin-Chave A, Cortet J, Hedde M, Capowiez Y. How agroforestry systems influence soil fauna and their functions—A review. Plant Soil. 2020;**453**(1):29-44. DOI: 10.1007/s11104-019-04322-4

[103] Guillot E, Bertrand I, Rumpel C, Gomez C, Arnal D, Abadie J, et al. Spatial heterogeneity of soil quality within a Mediterranean alley cropping agroforestry system: Comparison with a monocropping system. European Journal of Soil Biology. 2021;**105**:103330

[104] Lorenz K, Lal R. Soil organic carbon sequestration in agroforestry systems: A review. Agronomy for Sustainable Development. 2014;**34**(2): 443-454

[105] van Noordwijk M, Coe R, Sinclair FL, Luedeling E, Bayala J, Muthuri CW, et al. Climate change adaptation in and through agroforestry: Four decades of research initiated by Peter Huxley. Mitigation and Adaptation Strategies for Global Change. 2021; **26**(5):1-33

[106] Taylor NG, Grillas P, Al Hreisha H, Balkız Ö, Borie M, Boutron O, et al. The future for Mediterranean wetlands: 50 key issues and 50 important conservation research questions. Regional Environmental Change. 2021;**21**(2):1-17

[107] McCann S, Cazelles K, MacDougall AS, Fussmann GF, Bieg C, Cristescu M, et al. Landscape modification and nutrient-driven

instability at a distance. Ecology Letters. 2021;**24**(3):398-414

[108] Darrah SE, Shennan-Farpón Y, Loh J, Davidson NC, Finlayson CM, Gardner RC, et al. Improvements to the Wetland Extent Trends (WET) index as a tool for monitoring natural and human-made wetlands. Ecological Indicators. 2019;**99**:294-298

[109] FAO. Unasylva 251: Forests: Nature-based Solutions for Water. Rome, Italy: Food & Agriculture Org.; 2019. DOI: 104060/CA6842EN

[110] Kayranli B, Scholz M, Mustafa A, Hedmark A. Carbon storage and fluxes within freshwater wetlands: A critical review. Wetlands. 2010;**30**:111-124

[111] Page SE, Baird AJ. Peatlands and global change: Response and resilience. Annual Review of Environment and Resources. 2016;**41**:35-57

[112] UN Water. Available from: Scarcity | UN-Water (unwaterorg). [Accessed: October 14, 2021]

[113] Mukharamova S, Saveliev A, Ivanov M, Gafurov A, Yermolaev O. Estimating the soil erosion cover-management factor at the European part of Russia. ISPRS International Journal of Geo-Information. 2021;**10**(10):645

[114] Kurmangozhinov A, Xue W, Li X, Zeng F, Sabit R, Tusun T. High biomass production with abundant leaf litterfall is critical to ameliorating soil quality and productivity in reclaimed sandy desertification land. Journal of Environmental Management. 2020;**263**: 110373

[115] Ayangbenro AS, Babalola OO. Reclamation of arid and semi-arid soils: The role of plant growth-promoting archaea and bacteria. Current Plant Biology. 2021;**25**:100173

[116] Li C, Li Y, Ma J, Wang Y, Wang Z, Liu Y. Microbial community variation and its relationship with soil carbon accumulation during long-term oasis formation. Applied Soil Ecology. 2021;**168**:104126

[117] El-Sheekh M, Abdel-Daim MM, Okba M, Gharib S, Soliman A, El-Kassas H. Green technology for bioremediation of the eutrophication phenomenon in aquatic ecosystems: A review. African Journal of Acquatic Science. 2021;**46**(3):274-292. DOI: 10.2989/16085914.2020.1860892

[118] Mosa A, Taha AA, Elsaeid M. In-situ and ex-situ remediation of potentially toxic elements by humic acid extracted from different feedstocks: Experimental observations on a contaminated soil subjected to long-term irrigation with sewage effluents. Environmental Technology Innovation. 2021;**23**:101599. DOI: 101016/ jeti2021101599

[119] Ahmed DAEA, Gheda SF, Ismail GA. Efficacy of two seaweeds dry mass in bioremediation of heavy metal polluted soil and growth of radish (*Raphanus sativus* L) plant. Environmental Science and Pollution Research. 2021;**28**(10):12831-12846

[120] Bhattacharyya SS, Adeyemi MA, Onyeneke RU, Bhattacharyya S, Faborode HFB, Melchor-Martínez EM, et al. Nutrient budgeting—A robust indicator of soil–water-air contamination monitoring and prevention. Environmental Technology & Innovation. 2021;**24**:101944. DOI: 10.1016/j.eti.2021.101944

[121] Heinze C, Blenckner T, Martins H, Rusiecka D, Döscher R, Gehlen M, et al. The quiet crossing of ocean tipping points. Proceedings of the National Academy of Sciences. 2021;**118**(9):1-9. e2008478118. DOI: 101073/ pnas2008478118

[122] Sampaio E, Santos C, Rosa IC, Ferreira V, Pörtner HO, Duarte CM,

et al. Impacts of hypoxic events surpass those of future ocean warming and acidification. Nature Ecology Evolution. 2021;**5**(3):311-321

[123] Adamovic M, Al-Zubari W, Amani A, Ameztoy Aramendi I, Bacigalupi C, Barchiesi S, et al. Position paper on water, energy, food and ecosystem (WEFE) nexus and sustainable development goals (SDGs). In: Carmona Moreno C, Dondeynaz C, Biedler M, editors. EUR 29509 EN. Luxembourg: Publications Office of the European Union; 2019. ISBN 978-92-76-00159-1. DOI: 102760/31812, JRC114177

[124] Malagó A, Comero S, Bouraoui F, Kazezyılmaz-Alhan CM, Gawlik BM, Easton P, et al. An analytical framework to assess SDG targets within the context of WEFE nexus in the Mediterranean region. Resources, Conservation and Recycling. 2021;**164**:105205

[125] Nasrollahi H, Shirazizadeh R, Shirmohammadi R, Pourali O, Amidpour M. Unraveling the water-energy-food-environment nexus for climate change adaptation in Iran: Urmia Lake Basin case-study. Water. 2021;**13**(9):1282

[126] Botai JO, Botai CM, Ncongwane KP, Mpandeli S, Nhamo L, Masinde M, et al. A review of the water–energy–food nexus research in Africa. Sustainability. 2021;**13**(4):1762

[127] Marengo JA, Galdos MV, Challinor A, Cunha AP, Marin FR, Vianna MDS, et al. Drought in Northeast Brazil: A review of agricultural and policy adaptation options for food security. Climate Resilience and Sustainability. 2021:20. DOI: 101002/cli217. https://rmets.onlinelibrary.wiley.com/doi/full/10.1002/cli2.17

[128] Das K. Towards a smoother transition to organic farming. Economic and Political Weekly. 2007;**42**:2243-2245

[129] Tu C, Louws FJ, Creamer NG, Mueller JP, Brownie C, Fager K, et al. Responses of soil microbial biomass and N availability to transition strategies from conventional to organic farming systems. Agriculture, Ecosystems Environment. 2006;**113**(1-4): 206-215

[130] FAO and UNEP. Global Assessment of Soil Pollution: Report. Rome: FAO and UNEP; 2021. DOI: 104060/cb4894en

[131] Gunstone T, Cornelisse T, Klein K, Dubey A, Donley N. Pesticides and soil invertebrates: A hazard assessment. Frontiers in Environmental Science. 2021;**9**:122

[132] Riedo J, Wettstein FE, Rösch A, Herzog C, Banerjee S, Büchi L, et al. Widespread occurrence of pesticides in organically managed agricultural soils—The ghost of a conventional agricultural past? Environmental Science Technology. 2021;**55**(5):2919-2928

[133] Chowdhury A, Pradhan S, Saha M, Sanyal N. Impact of pesticides on soil microbiological parameters and possible bioremediation strategies. Indian Journal of Microbiology. 2008;**48**(1): 114-127

[134] Gujre N, Soni A, Rangan L, Tsang DC, Mitra S. Sustainable improvement of soil health utilizing biochar and arbuscular mycorrhizal fungi: A review. Environmental Pollution. 2021;**268**:115549

[135] Njeru EM. Crop diversification: A potential strategy to mitigate food insecurity by smallholders in sub-Saharan Africa. Journal of Agriculture, Food Systems, and Community Development. 2013;**3**(4):63-69

[136] Baudron F, Govaerts B, Verhulst N, McDonald A, Gérard B. Sparing or sharing land? Views from agricultural scientists Biological Conservation. 2021;**259**:109167

[137] FAO. In: Bélanger J, Pilling D, editors. The State of the World's Biodiversity for Food and Agriculture. Rome: FAO Commission on Genetic Resources for Food and Agriculture Rome; 2019

[138] Sarkar D, Kar SK, Chattopadhyay A, Rakshit A, Tripathi VK, Dubey PK, et al. Low input sustainable agriculture: A viable climate-smart option for boosting food production in a warming world. Ecological Indicators. 2020;**115**:106412

[139] Mourad KA, Hosseini SH, Avery H. The role of citizen science in sustainable agriculture. Sustainability. 2020;**12**(24):10375

[140] Ebitu L, Avery H, Mourad KA, Enyetu J. Citizen science for sustainable agriculture–A systematic literature review. Land Use Policy. 2021;**103**: 105326

[141] Slimi C, Prost M, Cerf M, Prost L. Exchanges among farmers' collectives in support of sustainable agriculture: From review to reconceptualization. Journal of Rural Studies. 2021;**83**:268-278. DOI: 101016/jjrurstud202101019

[142] Bizikova L, Nkonya E, Minah M, Hanisch M, Turaga RMR, Speranza CI, et al. A scoping review of the contributions of farmers' organizations to smallholder agriculture. Nature Food. 2020;**1**(10): 620-630

[143] Bardhan K, York LM, Hasanuzzaman M, Parekh V, Jena S, Pandya MN. Can smart nutrient applications optimize the plant's hidden half to improve drought resistance? Physiologia Plantarum. 2021;**172**(2): 1007-1015

[144] Thomas JBE, Sinha R, Strand Å, Söderqvist T, Stadmark J, Franzén F, et al. Marine biomass for a circular blue-green bioeconomy?: A life cycle perspective on closing nitrogen and phosphorus land-marine loops. Journal of Industrial Ecology. 2021. DOI: 10.1111/jiec.13177. https://onlinelibrary.wiley.com/doi/10.1111/jiec.13177

[145] Jupp AR, Beijer S, Narain GC, Schipper W, Slootweg JC. Phosphorus recovery and recycling–closing the loop. Chemical Society Reviews. 2021;**50**: 87-101. DOI: 101039/D0CS01150A

[146] Alewell C, Ringeval B, Ballabio C, Robinson DA, Panagos P, Borrelli P. Global phosphorus shortage will be aggravated by soil erosion. Nature. Communications. 2020;**11**(1):1-12

[147] Rezaei H, Saadat S, Mirkhani R, Bagheri YR, Esmaeelnejad L, Nouri Hosseini M, et al. The state of soil organic carbon in agricultural lands of Iran with different agroecological conditions. International Journal of Environmental Analytical Chemistry. 2020:1-17. DOI: 101080/0306731920201786547

[148] Tabatabaei SH, Nourmahnad N, Kermani SG, Tabatabaei SA, Najafi P, Heidarpour M. Urban wastewater reuse in agriculture for irrigation in arid and semi-arid regions—A review. International Journal of Recycling of Organic Waste in Agriculture. 2020;**9**(2):193-220

Chapter 5

Farmer's Perception of Associates Non-Cocoa Tree's Leaf Litterfall Fertilizing Potential in Cocoa-Based Agroforestry System

Milie Lionelle Tsouga Manga, René Menoh A Ngon, Etienne Akoutou Mvondo, Eunice Ndo, Bidzanga Nomo and Zachée Ambang

Abstract

Investigations to assess farmer's perceptions on the fertilizing potential of associated trees species in cocoa agroforest of degraded forest ecology were carried out in southern Cameroon. The perception of the farmers was based on the ability of the trees to maintain or improve soil fertility of their farms. The verification of these perceptions was done through an evaluation of litter fall biomass nutrient content (N, P, K, Ca and Mg) of selected trees. The top 5 associates trees ranked by farmers was: *Milicia excelsa, Ceiba pentandra, Ficus mucuso, Asltonia boonei, Terminalia superba.* The chemical analysis of the leaf litter from the different tree species revealed a significant different between their chemical components. N appeared to have the highest concentrations varying from 2.82 to 5.57% with a mean value of 4.25 ± 1.065%, P had the lowest concentrations typically around 0.001%. The top 5 tree species based on the chemical analysis ranking were: *C. pentandra*, *M. excelsa*, *Eribroma oblungum*, *Alstonia boonei*, *Zanthoxylum heitzi.* Farmer's perceptions thou holistic, are not completely different from scientific finding. Therefore, they should be taken in consideration in management plans for cocoa- based systems in order to enhance their ecological and economic performance.

Keywords: Farmer's perception, associated trees, fertilizing potential, Cocoa, Agroforestry

1. Introduction

The cocoa tree (*Theobroma cacao* L.) belongs to the Malvaceae family and is native to the tropical rain forests of Central and South America [1]. The Germans first introduced cocoa in Africa through Ghana in 1857 and in Cameroon precisely through Victoria (Limbe) in 1886 [2]. The nutritional and pharmaceutical importance of cocoa makes it one of the main export products for certain tropical countries. In Cameroon, and particularly in the center region, Cocoa is still grown in the traditional Agroforestry way with the shade of some forests, fruits trees and oil palms [3–5]. These systems have been widely described in relation to the environmental and ecosystem services they provide, but much less regarding productivity related to their structure [4, 6, 7].

Soil fertility under tropics is mainly influenced by biological interactions, and trees are the main driver of these, as they provide the rough material to achieve these [8]. Bellow and above grounds interactions significantly influence the status of soil health through their rooting habit, but also through the decomposition and mineralization of the litter fall. Several studies have demonstrated that the reduction of nutrient and organic matter content in the soil is a serious threat for agricultural production and food security in many tropical countries [9, 10]. Research has been involved in this theme to understand the mechanisms of conservation and improvement of soil fertility by trees. Numerous studies have identified links between traditional knowledge of trees associated with cocoa agroforest and soil health (fertility), although some processes are difficult to codify [8, 11, 12].

Although not always recognized by agronomists, trees in cocoa-based agroforests have more uses for local farmers than just providing a suitable microclimate for cocoa trees [3, 8]. Some indigenous species are maintained in the system by local farmers for their fertilizing capacities through nutrient recycling. Such tree species most mentioned as positively influencing fertility of soils and/or having other desirable attributes in traditional land used as described by [8, 11] include species such as *Ceiba pentandra* described to have floral and leaf litter fall that improves soil upon decomposition in rainy season, gathers dew in the dry season and prevents soil from drying, woody parts decompose rapidly and add to fertility, soil around always wet; *Milicia excelsa* described to provides good shade, improve microclimate, leaf litter improve soil conditions, deep rooting habit, abundant leaf litter and high leaf litter decomposition rate and soil fertility value; *Alstonia boonei* have *a* deep rooting habit, tall tree, wide and open crown, intermediate leaf area, abundant leaf litter with high decomposition rate, provides good shade and maintains soil moisture, gathers dew/exudes water, fertility value; *Ficus mucuso and Ficus exasperate* described to possess deep rooting habit, tall tree, wide and open crown, big leaf area with high leaf litter decomposition rate, high fertility value etc. Yet, little information is available on the contribution of these species leaf litter nutrient to the productivity of soils under cocoa fields. It is therefore hypothesized that trees species in cocoa-based agroforests play a major role in the improvement of soil conditions, hence the productivity of the entire system.

Appropriate tree species selection based on nutrient cycling is a vital issue in agroforestry practices [13]. So far, the screening or prioritization of commonly present indigenous trees species of cocoa-based agroforestry systems (CBAFS) is based on ethnoecological and ethnopedological studies [11, 12] but a scientific approach was carried base on the rate of mychorrizal colonization of the roots of some of these indigenous tree species. The results established a positive correlation between local farmer's classification of ten indigenous tree species with high fertility potential, and the colonization of the roots of those by mychorrizal [11]. However, no or little attempt has been taken to assess the fertility potential of these trees in term of their nutrient content present in their leaf litterfall, hence the need for this study.

This study, therefore, aims to investigates farmer's perceptions and knowledge on the fertilizing potential of the leaf litterfall of the non – cocoa tree associated to cocoa in CBAFS. If the pertinence of the perceptions is established, such measures could be introduced in management plans for cocoa- based systems in order to enhance their ecological and economic performance.

2. Methodology

2.1 Study site

This multiscale study encompassed the village area, the household and farming unit, the cocoa farm and homogeneous cocoa plantations. The study was conducted in the Centre Region, i.e., in the Evodoula subdivision (Latitude: 4° 04′ 60.00" N Longitude: 11° 11' 60.00" E) (**Figure 1**). Located in one of the oldest cocoa producing basin in a degraded forest zone, the Evodoula village (≈76 Pop/km^2) is characterized by a high land use intensity.

The climate is hot and humid with average temperatures and relative humidity of 25°C and 75% respectively. The rainfall pattern is bimodal with a heavy rainy season running from mid-September to mid-November, and a severe dry season running from December to mid-March.

Pedologically, there is a wide variety of soils based on structure and texture. The soils are mostly ferralitic, sandy clay and hydromorphic in the lowlands found around certain places. These soils are 90% agricultural land, favorable to the cultivation of cocoa and to a wide range of food and market garden products and are intensively exploited because of the strong demography. The operating method is based on clearing, cutting, burning and plowing, which helps to strip the vegetation cover and exposes the soil to severe erosion and reduced fertility. Unfortunately, the amendments (organic fertilizers) recommended by the competent services are used very little [14].

Vegetation is semi-deciduous evergreen and degrades from the gradient Equator to north [11].

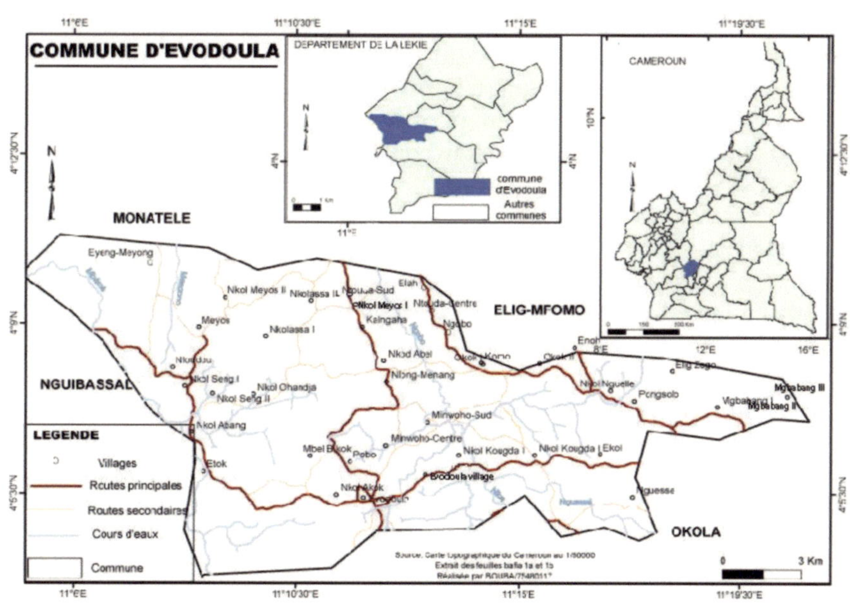

Figure 1.
Location of the study site in it agro-ecological zone.

2.2 Data collection

Both qualitative and quantitative data were collected from August 2013 to March 2014. Two different methods were used for data collection: (1) semi-structured socioeconomic surveys with households, and (2) direct observations and measurements in cocoa plantations. A total of forty (40) cocoa growers, were selected randomly for the semi-structured socioeconomic interview. These were focused on cocoa plantation characteristics (plantation status, age, history since its initial planting, area, cocoa tree ages, etc.) and the identification/selection and ranking of 10 tree species with leaf litterfall of high fertility potential was done. This identification/selection and ranking was based on their ethno-botanical knowledge of associated indigenous species, and the productivity of cocoa stands around those species. A generalized farmers' ranking was obtained by calculating the mean value of the position occupied in the individual farmers' ranking. The empirical classification by farmers of the fertility potential of these ten species was then compared to the classification of the same species based on their respective nutrient contents (Test Ranking).

Following these interviews with the cocoa growers, field visits to plantations were organized to select fifteen (15) cocoa agroforestry systems, through an in – depth assessment, for specific farm characterization. In each of the fifteen cocoa agroforestry systems, a systematic inventory of all non-cocoa trees exceeding 1 m in height were inventoried over the total area of each cocoa plantation following the method of [15]. Each tree species (forest, exotic as well as palm tree) was counted, numbered, identified and their density per plot estimated. The species identifications were based on vernacular names in the 'Eton' language with the assistance of the farm owner and correspondences with the scientific names were established from literature review [16].

From the above fifteen cocoa farms, litterfall of the 10 trees species of high fertility potential as rank by farmers was collected daily. Here, every newly fallen leaf was collected systematically at the same time after every two (02) days for one (1) week. The collection of fresh litterfall was done by a random walk around the specific tree species studied and the distance covered was from the base of the trunk to the longest branch of tree when the sun is at the zenith. Litterfall was conditioned according to standard procedure and taken to the laboratory for compositional analysis of macro-nutrients (nitrogen, phosphorus, potassium, calcium, and magnesium) and analyses were carried in conformity with standard analytical procedures of [17].

2.2.1 Data analyses

The data collected from the questionnaire and inventory forms were checked, entered into Microsoft Office Excel 2007 software and were analyzed using the Statistical Package for Social Sciences (SPSS) version 12.0. These analyzes consisted of descriptive statistics (sum, frequency, percentage, tree species densities and cross-tabulations of results), interactive graphs, and the total litter primary macro-nutrient (PMM) contents of the tree's species; which was obtained by summing the proportion of the respective elements analyzed. This, enable us to established the Test ranking of trees species. Data obtained from the chemical analysis were analyzed as a one-way analysis of variance (ANOVA) using the Proc GLM IN in SAS version 9.0. Separation of means was done using the DUNGAN Multiple Range Test, to test for significant effects between the leaf litters nutrient compositions of the different tree species at 5% probability level.

3. Results and discussions

3.1 Results

3.1.1 Cocoa farms characteristics

Many of the CBAFS monitored in the study area were cultivated and managed in very traditional ways resembling the approaches implemented previously by the elders. Among the households surveyed, 20% of the cocoa plantations had been established by the current owners, 80% had been established by grandparents or parents of the current owners (inherited). Those cocoa plantations had been established from food-crop fields (13%) or forest areas (73%). As a whole, 40% of the farms fall in the age range of (> 10–30 years), while 27% and 33% represent the age range of (≤ 10 years) and (> 30 years) respectively. Local classification of Cocoa cultivars used in the area identified two varieties; the local Cocoa landraces locally called "German variety" (80%) (**Figure 2**), mostly made up of Forastero-Amazonian or upper Amazon varieties and a considerable population of Trinitario in the old Cocoa orchard. Due to the fact that 40% of the plantation studied are old plantations that is >30-year-old, some of the farmers turn to regenerate their plantations consequently, some systems (13%) had a mixture of the "German cocoa" and the Hybrids while, the young systems (7%) were mostly dominated by the improved varieties or hybrids locally called 'SODECAO' from the name of the parastatal in charge of the distribution of the cultivars of that variety. Farmers had specific knowledge of the behavioral characteristics of each of cocoa cultivars.

3.1.2 Associate's tree species diversity

The results of the inventory of non-cocoa tree species diversity showed that there were in total 122 different non-cocoa trees species, with in total 1417 shade trees recorded over a total surface area of 29.5 ha, resulting to a shade tree density of 48 trees/ha of the different cocoa systems of the studied. Species sampled belonged to 37 tree families. The families mostly represented were: Sterculiaceae (28 species), Moraceae (22 species), Mimosaceae (15 species), Apocynaceae (13 species), Anacardiaceae (11 species), Euphorbiaceae (10 species), Meliaceae (10 species), Rutaceae (10 species), Bombaceae (8 species), Burseraceae (8 species) and Musaceae (8 species). Also, 493 trees belonging to 43 species were of the fruit types, while, 924 trees belonging to 79 specie were of the forest type. The 17 most occurring trees species are shown in **Table 1**.

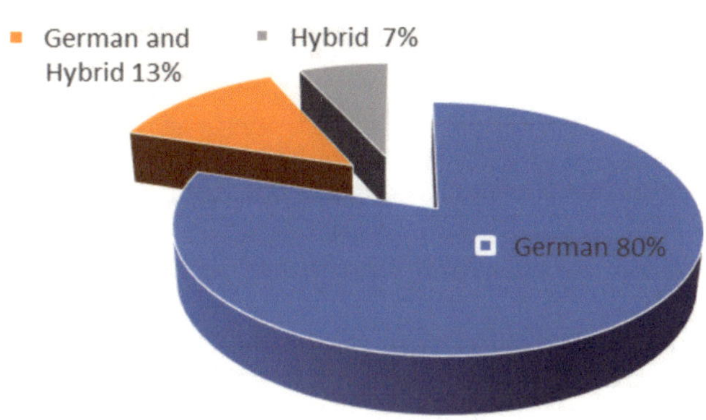

Figure 2.
Type of vegetative material used in the study zone.

No.	Scientific name	Local name	Family	Frequency	Percentage (%)
1	*Mangifera indica*	Andok ntangani	Anacardiaceae	135	18%
2	*Persea americana*	Pia	Lauraceae	131	17%
3	*Dacryodes edulis*	Sa'a	Burseraceae	81	11%
4	*Milicia excelsa*	Abang	Moraceae	66	9%
5	*Terminalia superba*	Akom	Combretaceae	64	8%
6	*Elaeis guineensis*	Allen	Arecaceae	59	8%
7	*Albizia adianthifolia*	Sayeme	Mimosaceae	37	5%
8	*Mansonia altissima*	Opong	Sterciliaceae	34	4%
9	*Citrus recticulata*	Opouma	Rutaceae	33	4%
10	*Ficus mucosa*	Ekokolle	Moraceae	24	3%
11	*Cola accuminata*	Abel	Sterculiacaeae	17	2%
12	*Musanga cecropoiides*	Eseng	Cecropiaceae	15	2%
13	*Lovoa trichilioides*	Bibollo	Meliaceae	14	2%
14	*Tieghemella africana*	Avo	Sapotaceae	13	2%
15	*Citrus sinensis*	Nia opouma	Rutaceae	13	2%
16	*Entandrophrama cylindricum*	Asse	Meliaceae	12	2%
17	*Nesogordonia papaverifera*	Kodibé	Sterculiacaeae	11	1%

Table 1.
Frequency and percentage of the 17 most common species present in the study location.

3.1.3 Farmers' perceptions and ranking of trees species indicators of fertile soils

Cocoa farmers interviewed confirm that, all cocoa-based systems studied (100%) were dominated by the presence of indigenous trees species. The latter produced more litter than the exotic trees species. The organic matter they produced played a favorable role in maintaining the soils' fertility. The farmers believe in their protecting ability against soil erosion (13%), maintenance and improvement of the soils' physico-chemical properties (structure; porosity; water retention, and soil nutrient) (100%), modification of the soil temperature (13%), and the rapid decomposition of organic matter (87%).

Farmers based their perceptions of the fertility potential of certain tree species on the observation of cocoa productivity around the tree (100%), vegetative aspect such as the size; the consistency and the arrangement in relation with the crown cover (93%). It is understood that, tree species introduced or maintained by farmers are those closer to the cocoa trees and through the quality of litterfall, shade, and eventual improvement of the soils' fertility plays an essential role on the cocoa productivity.

The ranking of the tree species with respect to their fertility potential, by local farmers, in decreasing order of importance is shown in **Table 2**.

3.1.4 Nutrient content of the litterfall of the rank tree species

Significant difference was found to exist in the chemical composition (quality) between litterfall from the different studied tree species ranked by farmers and within the different nutrient elements tested. Nitrogen was the main nutrient in leaf litter of different tree species with its concentration

No.	Trees species	Mean rank
1	*Milicia excelsa*	1.4
2	*Ceiba pentandra*	1.9
3	*Ficus mucoso*	2.8
4	*Alstonia boonei*	2.8
5	*Terminalia superba*	3
6	*Eribroma oblungum*	3.3
7	*Irvingia gabonensis*	4.2
8	*Zanthoxylum heitzi*	5.5
9	*Musanga cecropoides*	7.7
10	*Coula edulis*	9.9

Table 2.
Farmers mean ranking of tree species of indicators of fertile soil.

varying from 2.82 to 5.57% and a mean of 4.23 ± 1.065%. Phosphorus is present in very low concentration, typically around 0.001% while K varied widely from 1.95 to 18.9 cmol/kg. Mg was quantitatively the second element in leaf litter, with values ranging from 20 to 310.75 cmol/kg (**Table 3**).

Considering the importance of these nutrient elements to the growth of cocoa, its can be observe that, they are presence in very high concentration in the leaf litter of these trees' species. Concentration which are a way too far higher than the threshold values required for cocoa cultivation (**Table 4**).

Nutrient content species	N (%)	P (%)	K (cmol/kg)	Mg (cmol/kg)	Ca (cmol/kg)
Ceiba pentandra	5.57^{a}	0.17^{ab}	15.36^{abc}	108.25^{b}	50.25^{b}
Eribroma oblungum	5.36^{ab}	0.14^{abc}	16.36^{abc}	310.75^{a}	28.2^{cd}
Milicia excelsa	5.35^{ab}	0.192^{a}	15.62^{abc}	66.83^{bc}	6.78^{bc}
Alstonia boonei	4.95^{ab}	0.116^{abcd}	13.41^{c}	57.25^{bc}	33.9^{bc}
Zanthoxylum heitzi	4.58^{abc}	0.072^{cde}	18.90^{a}	60.17^{bc}	23.9^{bc}
Ficus mucoso	4.08^{bcd}	0.173^{ab}	17.49^{ab}	59.33^{bc}	168.35^{bc}
Musanga cecropoides	3.35^{cd}	0.033^{e}	5.26^{d}	244.5^{a}	29.2^{a}
Terminalia superba	3.27^{d}	0.109^{bcde}	13.90^{bc}	49^{bc}	13.7^{bc}
Irvingia gabonensis	2.98^{d}	0.087^{cde}	15.19^{abc}	20^{c}	60.55^{c}
Coula edulis	2.82^{d}	0.04^{de}	1.95^{d}	73.67^{bc}	16.45^{bc}
Anova/Duncan's Multiple Range Test					
Df	9	9	9	9	9
F value	6.92	5.28	22.09	17.15	30.71
Pr (>F)	0.0002	0.001	<.0001	<.0001	<.0001

Values in the same column followed by different superscript are statistically different at P < 0.05 level using Duncan's Multiple Range Test.

Table 3.
Chemical composition of the litter falls of the 10 trees species ranked by farmers.

Macronutrient	Unit	Threshold values
Total Nitrogen	%	0.2–0.4
Available Phosphorous	ppm	6.0–15.0
Potassium	Cmol + kg^{-1}	0.2–1.2
Calcium	Cmol + kg^{-1}	4.0–18.0
Magnesium	Cmol + kg^{-1}	0.9–4.0
Source: Snoeck et al., 2016.		

Table 4.
Average soil macronutrient threshold values for cocoa cultivation.

N^0	Local ranking	Ranking according to the sum of the PMN content	Sum of the PMN Content (g/kg)
1	*Milicia excelsa*	*Ceiba pentandra*	63.3
2	*C. pentandra*	*M. excelsa*	61.5
3	*Ficus mucoso*	*Eribroma oblungum*	61.3
4	*Alstonia boonei*	*Alstonia boonei*	55.9
5	*Terminalia superba*	*Zanthoxylum heitzi*	53.9
6	*Eribroma oblungum*	*Ficus mucoso*	49.4
7	*Irvingia gabonensis*	*T. superba*	39.3
8	*Zanthoxylum heitzi*	*Irvingia gabonensis*	36.6
9	*Musanga cecropoides*	*Musanga cecropoides*	35.8
10	*Coula edulis*	*Coula edulis*	29.4

Table 5.
Comparison of farmer's ranking (local) with the ranking from the sum of PMN content of the studied trees species.

3.1.5 Comparison between farmers' ranking and measured nutrient contents

Contribution of trees species in soil fertility sustenance in general was based on indicators such as cocoa productivity, abundance of biomass produced, functional attributes of certain organs of these species and interactions with the medium. The litterfall of the associated tree species, based on farmers' perceptions determine the biomass produced, which once decomposed, improves soil fertility. The above-mentioned criteria were the basis of the ranking of 10 species of high fertility potential in descending order of importance (**Table 1**) by farmers. Comparison of farmers ranking with the ranking obtained by summing the nutrient content of the primary macro-nutrients (PMN) (N, P and K) (Test ranking) is presented in **Table 5**. Farmers' ranking, though closer, but is different from the Test ranking.

3.2 Discussion

3.2.1 Characteristics of cocoa based agroforestry systems

This study was performed in an attempt to acquire farmers' perceptions of the fertility potential of associated non-cocoa trees in cocoa systems in order to develop more knowledge about the soil-trees interactions. The maximal farm size in the entire study area is 5 ha and the smallest cocoa fields have a surface area of 0.5 ha.

The smallest surface area observed in this zone is due to the fact that 60% of the cocoa agroforests are inherited. In this zone the beneficiaries share the heritage left by their parents or relatives. This factor further contributes to the reduction of the surface area of the plantation and does not facilitate the creation of new plantation because the pressure on the available land is high, these further account for the small number of cocoa agroforests of age range ≤ 10 years within the study area. These results are similar to those obtained by [18].

The fact that most cocoa agroforest ownership is acquired by inheritance could further explain the age of the cocoa agroforest. These results are closer to those found by [19]; [18] who confirms that Cocoa orchard of Center Cameroon are old from the fact that 70% of the cocoa farms are more than 40-year-old. These results practically indicate that farmers of the study zone do not create new plantations. This is due to the high pressure exerted on the available land, which does not facilitate the acquisition of land by the younger producers. This was also observed by [20, 21] who further noted that cocoa trees were not renewed in cocoa plantations and were as old as the plantations.

3.2.2 Non-cocoa trees species densities, frequencies and abundance

Cocoa based agroforestry systems (CBAFS) of the study area are complex multispecies cropping systems whose performances are usually difficult to assess. The associated non – cocoa tree species diversity was high, with a predominance of timber species. Nevertheless, the fruit tree species *Mangifera indica*, *Persea American* and *Dacryodes edulis* were the most occurring species. Located in the lekié division, near Yaoundé urban city this further confirm the finding of [22, 23], who demonstrated in their study that farmers introduced and maintained fruit trees in plantation for the sake of income diversification. The high abundance of non-primary forest species points to the degree of alteration of the cocoa agroforests compared to primary forest. These cocoa agroforests have a high tree diversity compared to cocoa production systems in other parts of the tropics. For instance, they are higher than those of 38 species in traditional CBAFS in Central Cameroon [24]. Results of this inventory are similar to those of 38 species in traditional CBAFS in Central Cameroon [24], 21 species identified [25] by and those of 40 species identified in three CBAFS (traditional, innovative and SODECAO) in the locality of Talba (Center Region of Cameroon) by [26]. The differences in species obtained could be explained by the fact that farmers in localities such as Talba established CBAFS following recommendations by SODECAO for cocoa cultivation and maybe by the smaller size of their sampling units.

3.2.3 Farmers ranking of non-cocoa trees of high fertility potential and the nutrient composition of their leaf litterfall

Farmers were able to identify and rank non-cocoa trees species they considered of having a high fertility potential of CBAFS in center Cameroon. These results joined those obtained by [8, 12] in their works, who identified farmers' preferred trees species as far as soil fertility maintenance is concerned. Farmers' ranking of trees consider as indicators of fertile soil in our study though closer but was different from the ranking obtained by [8] in his study on mycotrophy and farmer knowledge of tree species of high fertility potential in cocoa-based agroforest of southern Cameroon. Our results are also in line with the results of [27], through his work on a look at activities on preferred trees in farming systems of the main cocoa producing countries in Africa, also identified and ranked several species as preferred trees for cocoa cultivation by farmers.

The results showed that, there is a significant difference in chemical composition (quality) between leaf litterfall from the different selected tree species and within the different nutrient elements tested. These results also indicate that Nitrogen is the main nutrient in the litterfall of the different tree species concentration varying from 2.82 to 5.57% with a mean data of 4.25 ± 1.065%. Mg is quantitatively the second element in the leaf litterfall of the different tree's species studied. The studied nutrient element where present in very high concentration and and far above the critical value needed for cocoa growth. This is in line with the finding of [28], who stated that a large part of some of these nutrient's elements is found in the vegetation. For instance, it was found that, for Cameroon, N in the litter was about twice the amount removed by the yield, whereas for Malaysia, this ratio was nearly 5 [28].

3.2.4 Comparison between farmers' ranking and measured nutrient contents

Based on the total primary macronutrient content potentially released by associated species, it can be observed that *Milicia excelsa, Ceiba pentendra, Eribroma oblungum, Asltonia boonei* and *Zanthoxylum heitzi* contained the highest nutrient concentration can be considered the best trees species in terms of fertility potential. However, with phosphorus (P) being an essential plant nutrient contributing to the development of fruits (increases flowering), we can say that *M. excelsa* and *Ceiba pentendra* are good sources for the improvement of soil fertility status (good indicators of fertile soil) under cocoa based systems in the study locality.

Compared to farmers' ranking, these two species appear in the 1st and 2nd positions respectively, meanwhile in the test ranking they appear in the 2nd and 1st positions respectively. The order of the primary macronutrient concentrations and returns to the soil through litterfall as observed in the isolated tree is N > K > P while that of the secondary macronutrient is Mg > Ca (an indication of the ranges of nutrient elements in concentrations and returns via litterfall). These results are close to those of [29], working on nutrient stocks, nutrient cycling, and soil changes in cocoa agroforestry. It could therefore be deduced that nutrient concentration in the litterfall of some trees is higher compared to other tree species, consequently some trees have a high fertility potential compared to others. However, in the light of the differences observed between the various rankings: farmers' ranking, test ranking, and the nutrient content of the associated species could not, on its own, serve as a tool for validating farmers' perceptions. Other factors such as the rate of mychoryzal colonization of roots of associated species [11], and soil fauna activities are also known to be important drivers of the biological fertility of the soil and soil health in general [30].

4. Conclusion

The present study which aimed at identifying and classifying 10 top species of good fertilizing potential and then collect and analyze the litter fall from these species trees in other to bring out a link between farmers knowledge and scientific knowledge aimed at, enhancing system sustainability and productivity. An important correspondence was found between the farmers' ranking and the chemical content of the litterfall, supporting the assertion that the integration of local knowledge in global science may contribute to easily understand the above and below grounds interactions in agroforestry systems in general and therefore pave the way for a smooth adoption among end users. Considering the increasing climate change and the predicted negative impact on cocoa production in West Africa, this approach can be a subsequent widespread call for the adoption of climate smart cocoa production.

Acknowledgements

The authors would like to thank the project C2D/PAR/SAF: 'PROGRAMME D'APPUI A LA RECHERCHE AGRONOMIQUE DES SYSTEME AGROFORESTIERS' for the financial support. We are also grateful to farmers in the study locations for their interest and willingness to participate in the study.

Conflict of interest

"The authors declare no conflict of interest."

Author details

Milie Lionelle Tsouga Manga[1,2]*, René Menoh A Ngon[1,2], Etienne Akoutou Mvondo[1,2], Eunice Ndo[1], Bidzanga Nomo[1] and Zachée Ambang[2]

1 Institute of Agricultural Research for Development, IRAD, Yaoundé, Cameroon

2 Faculty of Science, Department of Plant Biology, University of Yaoundé I, Yaoundé, Cameroon

*Address all correspondence to: milielionnelle@ymail.com

References

[1] Alverson W. S., Whitlock B. A., Nyffeler R., Bayer C. & Baum D. A., 1999. Phylogénie du noyau de Malvales: prevue à partir des données de séquence *ndh* F. Amer. J. Bot., 86, (10): 12474-11486

[2] Voula V. A., Manga E. F., Messi A. L. M., Mahop R. & Begoude B. A. D., 2018. Impact des mirides et des infestations fongiques sur le dépérissement du cacao au Cameroun. J. Ent. Zoo. Stud., 6(5): 240-245.

[3] Sonwa D. J., Weise S. F., Schroth G., Janssens Marc J. J. & Sappiro H., 2014. Plant diversity management in cocoa agroforestry systems in West and Central Africa effects of markets and household needs. Agrofor. Syst., 88(214):1021-1034

[4] Wessel M. & Quist-Wessel P. M. F., 2015. Cocoa production in West Africa, a review and analysis of recent developments. *NJAS – Wag. J. Life Sci.*, 74-75, 1-7. doi.org/10.1016/j.njas.2015.09.001.

[5] Jagoret P., Dounias M. I. & Malézieux E., 2011. Long-term dynamics of cocoa agroforests: a case study in central Cameroon. Agrofor. Syst., 81:267-278

[6] Essola Etoa Louis Childéric., 2014. Evaluation des rendements potentiels en cacao (theobroma cacao l) dans les systèmes agroforestiers complexes en zone forestière à pluviométrie bimodale du centre cameroun. [Master dissertation]. Dschang: Faculty of Agronomy and Agricultural sciences, University of Dschang.

[7] Dagar J.C. & Tewari J.C., (eds) 2016. Agroforestry research developments. Nova Science Publishers, Inc, New York, p 578

[8] Bidzanga N., 2005. *Farmers' ecological and agronomic knowledge about the management of multi-strata cocoa systems in Southern Cameroon*. PhD thesis, School of Agriculture and Forest Sciences, University of Wales, UK p. 278.

[9] Mahmood H., Siddique M.R.H., Rahman M.S., Hossain M.Z. & Hasan M.M., 2011. Nutrient dynamics associated with leaf litter decomposition of three agroforestry tree species (*Azadirachta indica*, *Dalbergia sissoo* and *Melia azadirachta*) of Bangladesh, Journal of Forestry Research, 22, 577-582

[10] Saj S., Jagoret P., Todem Ngogue H. 2013. Carbon storage and density dynamics of associated trees in three contrasting *Theobrama cacao* agroforests of Central Cameroon. Agrofor. Syst. 87(6):1309-1320

[11] Fotsing B., 2008. *Perception paysanne de la Contribution des essences associées à la fertilité des sols dans les agroforêts à base de Cacaoyer en zone de transition forêt savane et de foret humide au sud Cameroun*. Master of science. Faculty of Science, University of Yaoundé I, Cameroon. 64 pp.

[12] Evans D., 2010. *Conversion of natural forest to cocoa agroforest in lowland humid Ghana: Impact on plant biomass production, organic carbon and nutrient dynamics*. PhD thesis, Kwame Nkrumah University of Science and Technology, Ghana, p 279.

[13] Hasanuzzaman M. d. & Mahmood H., 2014. Nutrient Return through Leaf litter Decomposition of Common Cropland Agroforest Tree Species of Bangladesh. *Int. Res. J. Biological Sci.*; 3(8), 82-88.

[14] United councils and cities of Cameroon, 2014. Administrative map of Cameroon. Available at: http://www.cvuc-uccc.com/national/index.php/en/

administrative-map/region-du-centre-2/90-association/carte-administrative/275-ngaoundal-14 [Accessed on: 2014-03-21]

[15] Aké A. L., 2002. *Flore de Côte d'Ivoire Catalogue Systématique, Biogéographie et Ecologie* (tome 2). Conservatoire et Jardin Botaniques: Genève, Suisse; 401p.

[16] Vivien J., & Faure J. J., 1985. *Arbres des forêts denses d'Afrique centrale*. Agence de coopération culturelle et technique, Paris ISBN 2-11-084, pp 796-4.

[17] Anderson J. M. & Ingram J., 1994. Tropical soil biology and fertility: a handbook of methods. Soil Sci 157 :265

[18] Kwesseu Petguen J.M., 2009. Analyse qualitatives des systèmes de cacao culture du Centre Cameroun. Mémoire de fin d'étude (diplôme d'ingénieur agronome). Faculté d'agronomie et des sciences agricoles. Université de Dchang. 133p

[19] Losch B., Fusillier J.L. & Dupraz P., 1991. Stratégies des producteurs en zone caféière et Cacaoyère du Cameroun. Quelles adaptations à la crise? Montpellier, France, Cirad–DSA, 252 p. (Coll. Documents systèmes agraires, n° 12).

[20] Michel I, Carriere S.M, Manga E. F, Bihina M. A, Blanchet A., Moisy C., Ngono F. & Levang P, 2019. Chapitre 6: Les cacaoyeres agroforestieres au Centre et au Sud Cameroun: diversité et dynamique. In: J. Seghieri (Coord.), J-M Harmand (Coord.) Agroforesterie et services écosystémiques en zone tropicale, Recherche de compromis entre services d'approvisionnement et autres services écosystémiques. SAFSE. Edition QUAE, Sciences update and Technologies, pp 85-98

[21] Manga E. F., Michel I., Mala W. A., Levang P., Ambang Z., Begoude B. D., Moisy C., Ngono F. & Carriere M. S., 2020. Cocoa-based agroforestry system dynamics and trends in the Akongo Sub region of central Cameroon. *Agrofor. Syst*. Doi.org/ 10.1007/ s10457-020-00510-9

[22] Jagoret P., Kwesseu J., Messie C., Dounias M. I. & Malézieux E., 2014. Farmer's assessment of the use value of agrobiodiversity in complex cocoa agroforestry systems in central Cameroon. *Agrofor. Syst.*, 88(6):983-1000. /doi.org/10.1007/ s10457-014-9698-1

[23] Rimlinger A., Carrie're S.M., Avana M.L., Nguegang A., Duminil J., 2019. The influence of farmers' strategies on local practices, knowledge and varietal diversity of the safou tree (*Dacryodes edulis* (G Don) H.J. Lam) in Western Cameroon. Econ Bot 73(2):249-264

[24] Madountsap T. N., Zapfack L., Chimi D. C., Kabelong B. L.P., Forbi P. F., Tsopmejio T. I., Tajeukem V. C., Ntonmen Y. A. F., Tabue M. R. B. & Nasang J. M., 2018. Carbon storage potential of cacao agroforestry systems of different age and management intensity, Climate and Development, DOI: 10.1080/17565529.2018.1456895

[25] Madountsap, T. N., Zapfack, L., Chimi, D. C., Kabelong, B. L. P., Tsopmejio, T. I., Forbi, P. F., ... & Nasang, J. M., 2017. Biodiversity and carbon stock in the SODECAO agroforestry system of center region of Cameroon: Case of talba locality. American Journal of Agriculture and Forestry, 5(4), 121-129.

[26] Mbolo A. M., Zekeng, J. C., Mala, A. W., Fobane, J. L., Chimi, D. C., Nyako, M. C., & Tamandjong, V., 2016. The role of Cocoa agroforestry in conserving forest tree diversity in Central Region of Cameroon. Agroforestry. Systems. 90 (4): 577-590

[27] Asare R., 2005. Cocoa agroforests in West Africa: a look at activities on preferred trees in the farming systems. *Forest and Landscape Working Papers no.* 6. ISBN: 87 – 7903 – 191 – 9

[28] Ndakara O. E., 2012. Litter fall and nutrient returns in isolated stands of *Terminalia catappa* trees in the rainforest area of southern Nigeria. *Ethiopian Journal of Environmental Studies and Management Vol. 5 No. 1 2012.* DOI: http://dx.doi.org/10.4314/ejesm.v5i1.1

[29] Hartemink A.E., 2005. Nutrient stocks, nutrient cycling, and soil changes in cocoa Ecosystems: a review. Adv Agro 86: 227 – 253

[30] Birang A.M., Victor A. & Bidzanga N. 2009. Assessment of some important fertility parameters related to different land use systems in the humid forest zone of southern Cameroon. *Int. J. Biol. Chem. Sci.* 3(1): 27-37. ISSN 1991-8631 Available online at http://www.ajol.info

Chapter 6

Vermicompost for Sustainable Agriculture and Bioconversion of Terrestrial Weed Biomass into Vermicompost

Chaichi Devi and Meena Khwairakpam

Abstract

Organic fertilizers are alternative to chemicals used in agriculture which enhance soil quality, prevent harmful chemicals entering into food chain, improve health and contribute to sustainable future socially, economically and ecologically. Vermicompost is a nutrient-rich organic fertilizer which promotes plant growth and improves soil quality. Vermicomposting is an economically feasible and environment friendly technology in which organic wastes are bio-converted into value added product and various organic wastes are used in this process. Terrestrial weeds are the plant species which grow on land and invasive in nature. These plants are responsible for various nuisances in the environment, agriculture and society. The weed biomass generated after various management methods are considered as organic waste. The terrestrial weed biomass is a possible option for the production of vermicompost. In this chapter scope of vermicompost for sustainable agriculture, the vemicomposting mechanism and the bioconversion of terrestrial weed biomass into vermicompost have been discussed.

Keywords: Organic fertilizer, Vernicompost, Sustainable agriculture, Terrestrial weed biomass

1. Introduction

Chemical fertilizers are used for a long period for nutrient supply in soil but inadequate amount and prolonged used of these chemicals proved to be detrimental for the environment and whole of ecosystem. Beneficial soil microorganisms as well as other biota have been disturbed due to continuous applications of chemical fertilizers. Therefore the needs of environment friendly and ecologically sustainable alternative have been realized. Organic fertilizers are always considered more sustainable practice of agriculture rather than chemical fertilizers [1]. Organic fertilizers have many advantages over chemical fertilizers. The demands for food is very high in current time with rapid increase of population all over the world specially the developing countries. The non-availability of land for growing crops led to the higher demand for application of chemical fertilizers which increased the crop production to fulfill the demand. But various adversities are associated with these chemical fertilizers and have shown direct impacts on soil, water and air. The runoff from agricultural areas caused eutrophication in nearby water bodies. The properties of soil are dramatically altered by the application of these chemicals, Rapid depletion of soil nutrients and soil water retention capacities are evident from various studies [2].

Organic farming relies on efficient utilization of local resources, and application of advanced sustainable technologies. The various management practices adopted in organic farming helps in soil improvement and better nutrient supply. The sustainable development to secure future generation is based on the concept of organic farming. Organic farming has various potential including soil fertility re-storage, sustainable agricultural produce, biodiversity mainatenance, food security and economic benefit to the marginal farmers. The research is oriented towards how organic farming is a solution to global food security and sustainable development. The international organizations are involved for policy development for cost-effective and sustainable strategies for stable food production. The major challenge is to shifting towards organic supplements and curbs the chemicals used in agriculture for a better and healthy environment [3]. Vermicomposting is a technology for the production of organic manure with the help of earthworms. Organic matters are degraded biologically resulting into a more stable compound called vermicompost rich in nutrients. There are various benefits of application of vermicompost to the soil as well as environment [4]. Various organic matters are utilized for the production of vermicompost and further research is going on to investigate the potential of various natural organic resources in the production of vermicompost [5–10]. Weeds are those unwanted plants that grow on any ecosystem disturbing the native plants and environment. Weeds which grow on land are termed as terrestrial weeds. Various management practices are adopted like mechanical, chemical, biological to remove and control these weeds in any forest or agricultural land. Terrestrial weed biomass is discarded as a waste after various management methods [11]. In the recent time utilization of terrestrial weed biomass for the production of vermicompost is documented in various researches [9, 11–14]. In the current chapter the role of organic fertilizers for sustainable development as an alternate superior solution to chemical fertilizers has been discussed. Various benefits of vermicomposting technology for the production of organic fertilizers and the potential of terrestrial weed biomass as a feedstock for the production of nutrient rich vermicompost is also illustrated.

2. Organic fertilizers as an alternative to chemical fertilizers

Heavy input of synthetic chemicals like herbicides, manure, pesticides in commercial agriculture includes ample amount of cash outpouring. Conventional agriculture practice always leads to pollution, soil degradation and health issues [15]. As a plant grows, the major essential seventeen elements along with other are taken up from the soil. The nutrient supply is the major role of soil and as these elements becomes in short supply than restoration is required in the soil for sustainable agriculture. Chemical fertilizers are used to replenish the soil for centuries. But the efficiency of these fertilizers is very low. Around 80–90% of the fertilizers are not assimilated by the plants after application to soil. The remaining fertilizers acts as contaminant and intrude into the environmental bodies [16]. The soil is capable to produce crops in sustainably but the application of various synthetic chemicals led to loss in soil fertility in different parts of the world. There is a huge gap between nutrient supply and nutrient demand. The nutrient in fertilizers was globally supplied in 2014 was 240 million tons while requirement of nutrient was 284 million tones. This results into scarcity in crop production [17]. On the other hand the heavy metals are present in chemical fertilizers and prolonged use of these results into adverse impacts on soil property. Plant metabolism is also affected by the accumulation of heavy metals and becomes fatal. The various contaminants from enters into food chain due to application of chemical fertilizers leading to bioaccumulation. The aquatic life of nearby water bodies are at high risk due to eutrophication [18]. Eutrophication is caused by runoff during rain and sometimes irrigation. Nitrogen and phosphorus are the major elements which mostly affect water bodies. Nitrogen in other forms like ammonia finds its way to atmosphere considered as one of the greenhouse gases. Negatively it contributes to global warming. Therefore all these chemicals are harmful to the surrounding ecosystem as well as human, plant and animal health. The mismanaged application of chemical fertilizers in excessive quantity often results into contamination. These contaminants accumulate in water; air and soil negatively affect the environment thus impedes sustainable production of food [19]. With growing demand the transition was seen in last decades from traditional practices to modern practices in agriculture. The prerequisite of inorganic fertilizers was very high wide-reaching to increase agricultural production. The stipulated demand increased for various essential elements in soil from 2015 to 2020. In the coming 5 years the overall fertilizer consumption will be very high. But eventual application of chemical fertilizers increases the chance of environmental contamination. These chemicals also contain heavy metals and other radioactive compounds becomes fatal when enters the food chain and also a major foundation for contaminants which exist in soil and environment for a longer period [16]. Generally by the composition the inorganic fertilizers more or less remain same as compared to organic fertilizers. However organic fertilizers always have advantages over chemical fertilizers both economically and ecologically. In the recent time amendment of soil with organic supplements is gaining tremendous response for the management of soil nutrients and sustainable agriculture [20].

Organic fertilizers are prepared from various organic resources and nutrient composition varies based on the characteristics of the used feedstock [11, 21, 22]. In the recent time on large scale production and application of organic fertilizers have been promoted worldwide. But the resources to prepare organic fertilizers are limited and difficult to meet the demand. Biodegradable organic matter is the major feedstock for organic fertilizer preparation. Organic matters are considered easiest way for nutrient recycling in soil and plants. They act as soil conditioner and dependency on chemical fertilizers is reduced. If the application is adequate with proper guidance to the farmers then small amount of organic fertilizers is enough to meet the requirement of essential elements for drop growth [16, 23]. Among the various advantages of organic fertilizers the major contribution is the supply of macro and micro nutrients, improvement in soil properties, increase in plant nutrient uptake efficiency, increase microbial interaction and also in a barren land it acts as a better amendment to revitalize infertility of the soil. Therefore a mass movement is required to shift in the application of organic fertilizers prepared from various organic sources. The application of chemical fertilizers as in conventional agricultural practices must be discouraged to avoid deterioration of food quality, human as well as soil health [16].

The nutrient level in organic fertilizers is very high. They are considered good soil amendments and also involves in various disease control in plants. Bacterial augmentation, promotion of plant growth, reducing population of nematodes has been observed during application of organic fertilizers. The increased microbial population and improved organic content in soil result into continuing reimbursement in soil fertility. The crop cultivation in organic way always retains stable pH and more exchangeable minerals for plant uptake [24]. The prevention of harmful impact of inorganic fertilizers and long-standing protection to land is possible by the application of organic manure. Organic fertilizers are generally free of toxic elements and heavy metals. The carbon linkages in these fertilizers make the nutrient ions release slower which consequently perks up the soil physical properties like aeration, water retention and sustainable nutrient supply. This helps in better plant growth [25].

3. Organic fertilizer for sustainable development

Organic food is in high demand due to consciousness for health among the consumers and concern for environment to meet sustainable development goals. Therefore most of the marginal growers are inclined towards organic farming. The research, promotion, marketing along with education and training for organic farming has been stimulated in various sectors all over the country. Organic farming includes good soil health, crop protection, organized plantation, health benefits and sustainable land use for agriculture. In both developed and developing nations organic farming plays a crucial role in maintaining healthy environment. The reduced level of greenhouse gas emissions, efficient energy use and protection of biodiversity as well as ecosystem in developed countries is the outcome of organic farming. On the other hand in developing countries it promotes high yield with economic benefits, sustainable utilization of resources and also maintains biodiversity [26]. The practical applicability of organic farming mostly depends on the farmers. Untill 2012 organic food shares US$ 60 billion of the global food industry. Organic farming is not about more nutrient supply to soil rather it concentrates to minimize loss of nutrients and proper soil management with sustainable nutrients supply. Retention and recycling of nutrients is the major goal of organic farming. Organic matters supply these nutrients in a healthy manner and prevent soluble nutrients to leach out [27]. The term sustainable development was first conceptualized in the Bruntland Commission Report in 1987. Sustainable development describes development as without compromising the future generation to fulfill their own needs while meeting the needs of the present generation. Keeping it in view, for the benefits of existing and future generation around 150 global leaders adopted a global agenda to achieve sustainable development by 2030 through 17 primary sustainable development goals (SDG) for transformation a better future. All these goals targets for an equitable society and mitigate climate change [17]. The SDG also includes zero hunger, good health and well being. Organic farming is vital to achieve these goals. The change in agricultural pattern is always suggested for food security and feeding ever expanding global population sustainably. Organic farming is always kept as topmost solution for this by various scientific reports [28–30]. According to [17] organic agriculture is a holistic approach for better soil quality, biodiversity, proper biological cycle and overall healthy agro-ecosystem. This can be accomplished by using any biological resources despite of synthetic chemicals. This is the base for organic movement throughout the world [31–33]. The principles lie on good health and ecology makes the root for organic farming to grow and gain popularity across the globe. Organic farming stands as a sustainable approach that reduces green house gas emission as a mitigation measure to climate change. Organic food production as a safety concern influences both consumer awareness and consumer purchase. In USA alone organic products shares 5.3% of the food market with $47 billion in 2016. In developing country like India organic market shares in 2015 $104.5 billion. Organic farming is always beneficial to the marginal farmers with who follows traditional approach with no chemical applications. This results into large scale organic production. On the other hand it provides economic benefits, health and environment, protection of biodiversity by the implication of traditional knowledge. The adaptation to climate change is possible through organic farming as it is concordant to traditional farming and more resistant to disease and extreme climate. The motto of SDG is to make the world poverty liberated and sustainable. This makes organic farming as a fundamental strategy for the well being of the planet. Organic farming is a combination of tradition and scientific innovation for the benefit of environment and maintaining a healthy relation of the organisms involved. This totally brings to an end of synthetic chemicals considered as a curse to environment [15]. In this context, organic fertilizer is a boon to organic farming as well as sustainable development [34, 35]. Organic matters are the major source for organic fertilizers. Addition of organic source to soil is always beneficial to maintain soil fertility. The various organic sources are still under investigation which can be utilized to prepare organic fertilizers. The green waste are considered biodegradable wastes and generated from various sources. The biomass of various organic wastes can be turned into wealth with proper technology to prepare organic fertilizers. In the following part of the chapter one of such technology and a possible source for organic fertilizer has been discussed in brief.

4. Vermicompost as nutrient rich organic fertilizer

Composting is a process in which biodegradation of organic matter takes place in an aerobic environment. There is various composting process and is one of the traditional practice for organic waste management. Over the time composting techniques have been evolved with various advanced engineering. Vermicomposting is an extension of composting technology in which earthworms are involved in the degradation process. Earthworms consume organic matter and accelerate degradation process. At the end nutrient rich vermicompost considered as one of the finest organic fertilizer is obtained [12]. Vermicompost is termed as black gold. Application of vermicompost has various benefits including promoting soil health and plant growth [36–38]. Being organic in nature is always superior to the other synthetic chemicals. The demand for vermicompost production is rising in global market. The outcome of vermicompost never harm human health and always safe for consumption. The environmental issues are always resolved by the application of vermicompost. The technology itself for vermicompost production is environment friendly and economical. The zero pollution and low cost of vermicomposting technology makes it advantageous over other fertilizer production technologies [11]. At present for a sustainable future adoption of an eco friendly and economical technology is very much important. Vermicomposting has the potential to mitigate all the problems related to health, environment and society. The organic waste generation is becoming a major problem all over the world and their ultimate fate is the waste stream. The burning also releases harmful elements to the atmosphere as well as destroys overall physical properties of the soil. The proper waste management through utilization of these bio wastes as organic fertilizer for organic farming is possible [39]. This may set a benchmark as the most significant approach for sustainable development.

The importance of earthworms for the biodegradation of organic waste was first observed by Charles Darwin which became the base for evolution of modern sustainable technology called vermicomposting for organic waste management [12]. Vermicomposting have evolved in more scientific way by different researchers [40–44]. There are various benefits associated with vermicompost. Vermicomposting is considered more advantageous than traditional composting process [11]. The particle obtained after breakdown during vermocomposting is more homogenous and uniform with earthy manifestation than heterogeneous mixture obtained during composting process [45]. The release of nutrients is slow during application of vermicompost to the soil. This benefits the easy uptake of nutrients by plants and also increases water holding capacity of soil [39]. Vermicomposting highly influence the soil physical properties. The soil porosity, aeration and temperature are well maintained during application of vermicompost. Soil microbial activity is also enhanced and replenishes nutrient content in soil in significant way which gradually promotes healthy plant growth [36]. Bulk density of soil is also observed to be reduced due to enhanced microbial activity resulting into increased porosity. The oxidation potential is also enhanced during vermicompost application to the soil. Vermicomposting of different organic wastes with different earthworm species results into nutrient rich eco-friendly vermicompost. The nutrient content is significantly high in vermicompost produced from various feedstocks and acts as a soil amendment for organic farming [9, 20]. The nutrients are released into more exchangeable format during stabilization process of organic waste making these elements readily available to plants in final vermicompost [12].

If we consider the economic benefits of vermicompost, in the current time global market for fertilizer is increasing. Farmers are more concentrated in efficient production. Organic fertilizers are always derived from natural sources. Vermicompost is basically a result of bio-conversion of various organic wastes. The improvement in soil quality and healthy crops associated with vermicompost application highly influence the rural economy. The economic profit for any agricultural produce in evaluated on the basis of yield, return and cost–benefit analysis. The nutrient losses from the field are marginalized during any organic application including vermicomposting. The lower emission of pollutants, soil biota conservation, enhanced nutrients is all characterized for economic profit for any marginal farmers using organic manure like vermicompost [20]. On the other hand on site production of vermicompost also requires very low input for installation and the raw material can be easily available without any transportation costs [12]. The rural development is possible through employment generation by small entrepreneur with vermicompost production with readily available resources in and around. The farmers were benefited substantially with organic farming as compared to equivalent economic input in synthetic fertilizers [20].

5. Terrestrial weed biomass as an organic waste for vermicompost

Various organic wastes have been investigated for the production of vermicompost. The unwanted plants in any ecosystem are termed as weeds and terrestrial weeds are those grow on land. The non native species to a particular area causing harm environment and health is termed as an invasive species. These invaders remain persistent in an area and gradually harm the total environment. The weeds particularly invasive in nature and destroys native vegetation. The whole nutrient regime and energy balance is altered by the prevailing of these weed species. The management of these weed species is difficult and complete eradication is not possible [12]. The weed biomass is a potential resource for the production of vermicompost. In the recent time various researches revealed the utilization of terrestrial weed biomass and found significant results while producing vermicompost [11–14]. Among the most noxious terrestrial weed species mostly investigated for the potential of vermicompost production are *Lantana camara, Ageratum conyzoides, Parthenium hysterophorus* etc. All of these weeds species are associated with various adversities and management with sustainable technology is very much required. The rapid regeneration capacity and morphologal adaptation make these weeds wide spread to a larger area. Most of these are found in forest, agriculture and urban ecosystem [12]. The weed biomass is easily available near farm area after clearance in abundance and can be easily utilized for onsite vermicompost production without any heavy cost involved. The nutrient rich vermicompost is produced from *Ageratum conyzoides* without any phytotoxicity. The germination index analysis proves *Ageratum conyzoides* biomass as an excellent media for plants [14]. The enhanced nutrient content in the bioconversion of *Ageratum conyzoides* biomass indicates its suitability as a substrate for vermicompost production [12]. The high tolerance capacity makes the weed *Lantana camara* to withstand any adverse environmental condition. Annually *Lantana camara* biomasss tons in billion is generated and can be utilized in an extensive way for vermicompost production. The FTIR analysis of vermicompost produced from *Lantana camara* biomass indicates reduced lignin content and allelopathic elements. Thus the vermicompost produced become eco-friendly and plant friendly [9].

6. Conclusion

Sustainable development depends on the conservation of natural resources and reliance on renewable resources for the future generation. Hunger reduction and food security for all is the major goal of SDG. The consumer consciousness for health and well being is the major driven force for organic movement. The chemicals applied in the paradigm of shift in agricultural activities in recent past have destroyed the food quality, soil fertility and the environment to a large extent. Organic farming is a solution to overcome all of these problems. The organic products are in high demand in both developed and developing nations. It was observed that organic farming is beneficial to the marginal farmers in rural economy and also helps in maintaining a clean environment. Organic fertilizers make organic farming more sustainable. Crop production, soil fertility is also enhanced due to application of organic fertilizers. The role of vermicompost as an organic fertilizer plays a crucial function in sustainable eco-friendly farming with abundant benefits including maintaining soil quality and nutrient supply to the crops in efficient way. Vermicompost application also accounts for pest control, enhanced microbial activity in soil and release of no emission to the atmosphere. Terrestrial weed biomass is seen to be a good resource for vermicompost production. This can be utilized in an efficient way instead of discarding it into waste stream. The proper management of terrestrial weed biomass through vermicomposting can resolve the demand for organic fertilizer globally.

Conflict of interest

The authors declare no conflict of interest.

Author details

Chaichi Devi[1]* and Meena Khwairakpam[2]

1 Department of Civil Engineering, National Institute of Technology Meghalaya, Shillong, Meghalaya, India

2 Centre for Rural Technology, Indian Institute of Technology Guwahati, Guwahati, Assam, India

*Address all correspondence to: chaichi.123@gmail.com

References

[1] Chang, E.H., Chung, R.S., Tsai, Y.H. Effect of different application rates of organic fertilizer on soil enzyme activity and microbial population. Soil Science and Plant Nutrition. 2007: 53(2):132-140. doi:10.1111/j.1747-0765.2007.00122.x

[2] Akram, M.S., Cheema, M.A., Nadeem, F., Waqas, M., Bilal, M., Saeed, M.. Role of bio-fertilizers in sustainable agriculture. Mediterranean Journal of Basic and Applied Sciences. 2020:4(2): 8-23.

[3] Kilcher, L. How Organic Agriculture Contributes to Sustainable Development. Journal of Agriculture and Rural Development in the Tropics and Subtropics. 2007:89

[4] Devi, C., Khwairakpam, M. Feasibility of vermicomposting for the management of terrestrial weed Ageratum conyzoides using earthworm species Eisenia fetida. Environmental Technology and Innovation. 2020:18.

[5] Ananthavalli, R., Ramadas, V., Paul, J.A., Selvi, B.K., Karmegam, N. Vermistabilization of seaweeds using an indigenous earthworm species, *Perionyx excavatus* (Perrier). Ecological Engineering. 2019:130:23-31.

[6] Das, D., Bhattacharyya, P., Ghosh B.C., Banik, P. Bioconversion and biodynamics of Eisenia foetida in different organic wastes through microbially enriched vermiconversion technologies. Ecological Engineering. 2016:86:154-161.

[7] Garg, P., Gupta, A., Satya, S. Vermicomposting of different types of waste using *Eisenia foetida*:A comparative study. Bioresource Technology. 2005.:97:391-395.

[8] Hanc, A., Chadimova, Z. Nutrient recovery from apple pomace waste by vermicomposting technology. Bioresource Technology. 2014:168:240-244.

[9] Hussain, N., Abbasi, T., Abbasi, S.A. Vermicomposting eliminates the toxicity of Lantana (Lantana camara) and turns it into a plant friendly organic fertilizer. Journal of Hazardous Materials. 2015. http://dx.doi.org/doi:10.1016/j.jhazmat.2015.04.073

[10] Khwairakpam M., Bhargava R. Vermitechnology for sewage sludge recycling. Journal of Hazardous Materials. 2009:161:948-954.

[11] Devi, C., Khwairakpam, M. Bioconversion of Lantana camara by vermicomposting with two different earthworm species in monoculture. Bioresource Technology. 2019:296. https://doi.org/10.1016/j.biortech.2019.122308

[12] Devi, C., Khwairakpam, M.. Management of lignocellulosic green waste Saccharum spontaenum through vermicomposting with cow dung. Waste Management. 2020:113.

[13] Devi, C., Khwairakpam, M. Management of invasive weed Parthenium hysterophorus through vermicomposting using a polyculture of Eisenia fetida and Eudrilus eugeniae. Environmental Science and Pollution Research. 2021. doi:10.1007/s11356-021-12720-4

[14] Gusain, R., Suthar, S. Vermicomposting of invasive weed Ageratum conyzoids: Assessment of nutrient mineralization, enzymatic activities, and microbial properties. Bioresource Technology. 2020:123537. doi:10.1016/j.biortech.2020.12353

[15] Setboonsarng, S., Gregorio, E. E. Achieving sustainable development goals through organic agriculture:

Empowering poor women to build the future. ADB Southeast Asia Working Paper Series, Mandaluyong City, Philippines: Asian Development Bank.2015:15

[16] Naeem, M., Ansari, A. A., Gill, S. S., (Eds.). Contaminants in Agriculture. 2020. doi:10.1007/978-3-030-41552-5

[17] FAO. Transforming Food and Agriculture to Achieve the SDGs. 20 interconnected actions to guide decision-makers. 2018. https://doi. org/10.1061/ (ASCE)MT.1943-5533. 0001328

[18] Lin, S.S., Shen, S.L., Zhou, A., Lyu, H.M. Assessment and management of lake eutrophication: A case study in Lake Erhai, China. Science of The Total Environment. 2020:141618. doi:10.1016/j.scitotenv.2020.141

[19] Bisht, N., Chauhan, P.S. Excessive and disproportionate use of chemicals cause soil contamination and nutritional stress. Soil Contamination - Threats and Sustainable Solutions. IntechOpen Book Series, 2020. DOI: 10.5772/intechopen. 94593.

[20] Lim, S. L., Wu, T. Y., Lim, P. N., Shak, K. P. Y. The use of vermicompost in organic farming: overview, effects on soil and economics. Journal of the Science of Food and Agriculture. 2014:95(6): 1143-1156. doi:10.1002/ jsfa.6849

[21] Ramnarain, Y. I., Ansari, A. A., Ori, L. Vermicomposting of different organic materials using the epigeic earthworm Eisenia foetida. International Journal of Recycling of Organic Waste in Agriculture. 2018:8(1):23-36. doi:10.1007/ s40093-018-0225-7

[22] Hussain, N., Abbasi, S., 2018. Efficacy of the Vermicomposts of Different Organic Wastes as "Clean" Fertilizers. State-of-the-Art. Sustainability. 10(4), 1205. doi:10.3390/ su10041205

[23] Sudharmaidevi, C. R., Thampatti, K. C. M., Saifudeen, N. Rapid production of organic fertilizer from degradable waste by thermochemical processing. International Journal of Recycling of Organic Waste in Agriculture. 2016:6(1):1-11. doi:10.1007/ s40093-016-0147-1

[24] Lee, J. Effect of application methods of organic fertilizer on growth, soil chemical properties and microbial densities in organic bulb onion production. Scientia Horticulturae. 2010:124(3):299-305. doi:10.1016/j. scienta.2010.01.004

[25] Shaji, H., Chandran, V., Mathew, L. Organic fertilizers as a route to controlled release of nutrients. Controlled Release Fertilizers for Sustainable Agriculture. 2021:231-245. doi:10.1016/b978-0-12-819555-0. 00013-3

[26] Yadav, S. K., Babu, S., Yadav, M. K., Singh, K., Yadav, G. S., Pal, S.A. Review of Organic Farming for Sustainable Agriculture in Northern India. International Journal of Agronomy. 2013:1-8. doi:10.1155/2013/718145

[27] Francis, C. A. Organic Farming Change History. Reference Module in Earth Systems and Environmental Sciences. 2013. doi:10.1016/ b978-0-12-409548-9.05237-4

[28] Pimentel, D., Hepperly, P., Hanson, J., Douds, D., Seidel, R. Environmental, Energetic, and Economic Comparisons of Organic and Conventional Farming Systems. BioScience, 2005:55(7):573-582. https://doi. org/10.1641/0006-3568(2005)055[0573,eeaeco]2.0.co;2

[29] Meemken, E.-M., Qaim, M. Organic Agriculture, Food Security, and the Environment. Annual Review of Resource Economics. 2018:10(1):39-63.

https://doi.org/10.1146/annurev-resource-100517-023252

[30] Reganold, J. P., Wachter, J. M. Organic agriculture in the twenty-first century. Nature Plants. 2016:2(2):15221. https://doi.org/10.1038/ nplants. 2015.221

[31] Das, S., Chatterjee, A., & Pal, T. K. (2020). Organic farming in India: a vision towards a healthy nation. Food Quality and Safety, 4(2), 69-76. doi:10.1093/fqsafe/fyaa018

[32] Obach, B.K. Organic Struggle: The Movement for Sustainable Agriculture in the United States. Food, Health and the Environment. The MIT Press, 2015

[33] Janick, J. Proceedings of the workshop on the history of the organic movement. 88th ASHS Annual Meeting, The Pennsylvania State University, University Park, 1991

[34] Meena, R. S. (Ed.). Nutrient Dynamics for Sustainable Crop Production. 2020. doi:10.1007/ 978-981-13-8660-2

[35] Ye, L., Zhao, X., Bao, E., Li, J., Zou, Z., Cao, K. Bio-organic fertilizer with reduced rates of chemical fertilization improves soil fertility and enhances tomato yield and quality. Scientific Reports. 2020:10(1). doi:10.1038/ s41598-019-56954-2

[36] Rekha, G. S., Kaleena, P. K., Elumalai, D., Srikumaran, M. P., Maheswari, V. N. Effects of vermicompost and plant growth enhancers on the exo-morphological features of Capsicum annum (Linn.) Hepper. International Journal of Recycling of Organic Waste in Agriculture. 2018:7(1):83-88. doi:10.1007/s40093-017-0191-5

[37] Blouin, M., Barrere, J., Meyer, N. Vermicompost significantly affects plant growth. A meta-analysis. Agronomy Sustainble Development. 2019:**39:** 34. https://doi.org/10.1007/s13593-019-0579-x

[38] Giri, B., Varma, A. (Eds.). Vermicompost and soil health. Soil Health. Soil Biology. 2020. doi:10.1007/978-3-030-44364-1

[39] Ganeshnauth, V., Jaikishun, S., Ansari, A. A., Homenauth, O. The Effect of Vermicompost and Other Fertilizers on the Growth and Productivity of Pepper Plants in Guyana. Automation in Agriculture - Securing Food Supplies for Future Generations. 2018. doi:10.5772/ intechopen.73262

[40] Dominguez, J. *Earthworms and vermicomposting.* Earthworms-The Ecological engineers of Soil. 2018. DOI: 10.5772/intechopen.76088.

[41] Dominguez, J., Aira, M., Kolbe, A.R., Gomez-Brandon, M., Perez-Losada, M. Changes in the composition and function of bacterial communities during vermicomposting may explain beneficial properties of vermicompost. Scientific Reports. 2019: 9. https://doi.org/10.1038/s41598-019-46018-w.

[42] Gajalakshmi, S. High-rate composting–vermicomposting of water hyacinth (*Eichhornia crassipes*, Mart. Solms). Bioresource Technology. 2002:83(3). doi:10.1016/s0960-8524(01) 00216-4.

[43] Huang, K., Xia, H., Cui, G., Li, F. Effect of earthworms on nitrification and ammonia oxidizers in vermicomposting systems for recycling of fruit and vegetable wastes. Science of the Total Environment. 2016:578.

[44] Cai, L., Sun, X., Hao, D., Li, S., Gong, X., Ding, H., Yu, K. Sugarcane bagasse amendment improves the quality of green waste vermicompost and the growth of *Eisenia fetida*. Frontiers of Environmental Science &

Engineering. 2020:14(4). https://doi.org/10.1007/s11783-020-1240-2.

[45] Lagcano, C., Dominguez, J. The use of vermicompost in sustainable agriculture: impact on plant growth and soil fertility. Soil Nutrients. 2011. ISBN 978-1-61324-785-3.

Chapter 7

Biosynthesis of Zinc Nanocomplex Employing for Plant Growth Promotion and Bio-Control of *Pythium ultimum*

Shaima M.N. Moustafa and Rania H. Taha

Abstract

Green Biosynthesis method was used for the preparation of Zn(II) nano complex from the reaction of the schiff base ligand 2,2′-((1E,1′E)-(1,2-phenylenebis (azanylylidene)) bis(methanylylidene))bis(4-bromophenol) and Zn(II)sulphate. The nano complex was characterized by different physicochemical methods. Zinc nanoparticles (ZnNP-T) will be studied as an antifungal agent. In this study, we will investigate the ability of the myogenic Zinc nanoparticles for plant Growth Promotion and Bio-control of *Pythium ultimum*.

Keywords: Zn-Nanocomplex, *Trichoderma harzianum*, *Pythium ultimum*, antifungal activity, *Trigonella foenum-graecum*

1. Introduction

The rising interest of consumers for food integrity and for the social and environmental sustainability of agriculture systems has a special impact on the crop production fields. This circumstance forces us to look for new tools for crop protection that are not based on chemical control.

Since the rising demand production of organic food, using of biofertilizers and biopesticides were the best way to alternative eco-friendly production. In fact, *Trichoderma* spp. have the greatest resistant to pesticides, which categorizes as integrated control and good bio-control agents [1, 2] because of its ability to produce many enzymes, such as chitinases, glucanases and proteases. In addition to bio-control, *Trichoderma* spp. have been recognized as plant growth promoter and promoters of different plant defense mechanisms [1, 3].

Nanoparticles have very important role in the biotechnology industries [4], which considered one of the fastest growing fields because of their biological, chemical and physical characterization. The biosynthesis of nanoparticles complex from fungi are a significant branch, due to the fungi has the greatest tolerance and metal bioaccumulation capability. Biogenic methods for nanoparticles forming by plants [5, 6], fungi [7, 8], and bacteria [9, 10] have formerly been described.

A great promise in this area is the Schiff bases. A Schiff base can be defined as the nitrogen analogue of aldehyde in which the carbonyl group is replaced by the azomethen group [11]. The transition metal complexes having O and N donor of Schiff bases ligands possess unusual configuration, structural liability, and are sensitive to the molecular environment [12].

Fenugreek (*Trigonella foenum-graecum* L) is an annual herb grown in North Africa, Egypt, India, Afghanistan, Bangladesh, Pakistan, Nepal, Iran, Turkey, Morocco, Spain, southern China and southern Europe. it is one of the commonly used herbs since ancient times. Medical Papyri of ancient Egyptian mentioned that it is used as food and antipyretic and in compounds used as incense for embalming and fumigation [13]. Several Arab writers described the plant and seeds as aperient, suppurative, emmenagogue, diuretic, useful in enlargements of spleen and liver, dropsy, and chronic cough. Fenugreek seeds contain 45.4% dietary fiber that blunts glucose absorption after a meal and regulate the production of cholesterol in liver [14, 15]. Seeds contain trypsine, alkaloids and chymotrypsine inhibitors [16, 17], anti-inflammatory steroidal saponin glycosides, flavonoids, furostanol steroidal saponins, flavone C-glycosides, spirostanol saponins and 17 amino acids, 7 of them being essential amino acids. Seeds considered as one of the most effective antidiabetic plants, used by traditional healers of northern Europe to treat diabetes, as stimulant and carminative, and in renal disorders. In Danish popular medicine, it is used to treat anxiety and depression. The seeds are used for wound dressing, rheumatism, stomachache, and leprosy, and have observed uterine stimulant activity [18].

However, its production may be affected by several exogenous stresses, including diseases caused by soil born pathogen microorganism. *Pythium* species affecting the different plants in younger stages and causing damping-off disease, this pathogenic fungi can caused root rot in later stages of plant growth [19, 20]. When the plant is in the seedling stage of its life cycle, infection by pathogenic *Pythium* species causes pre and post-emergence damping-off disease, which decaying the seeds and seedlings before and after the emergence of the plant from the soil surface, respectively. However, infected mature plants have also been found to show root rot symptoms.

Pythium spp. affect the young tissues, which have not developed to secondary thickenings; thus, the infection is limited to seedlings and youngest roots. The post-emergence damping-off disease of seedlings is associated with symptoms like water soaking, reduced growth, black or brown discoloration, wilting and root rot [21, 22]. In mature plants, water-soaked roots and lesions of stem at the soil line, stunted growth, and brown discoloration of roots are prevalent [23]. Due to their capability to disperse through different routes, their detection and Managements have become crucial. Therefore, the use of an integrated approach with bio and myogenic techniques for controlling of pathogenic *Pythium* spp. can be the most and best sustainable alternative to the traditionally used and dangerous chemical approach.

This study will focus on the management strategies of *Pythium ultimum* causing damping-off and root rot in Fenugreek (*Trigonella foenum-graecum* L) plants by preparation of Nano Schiff base complex, its characterization and its hopeful applications in different fields by using *Trichoderma harzianum*, which is a nonpathogenic, fast growing and environmentally friendly fungus. Therefore, the goals of the present study were: (a) to evaluate the pathogenicity capacity of nano metal complex against seedling. (b) to evaluate the tolerance of different concentrations of nano metal complex on the germination of seeds and possibility for acceleration of the growth and finally, (c) to evaluate the biosynthesis of nanoparticles complex for control of the disease caused by *P. ultimum* in seedlings.

2. Methodology

2.1 Materials

In this study, all chemicals used were of highest purity. Organic solvents used included C_2H_5OH and (DMF) were spectroscopic pure from British Drug House (BDH). Distilled water collected from glass equipment. The other materials such as, 5-bromo-salicylaldehyde (Sigma), *o*-phenelendiamine (Aldrich), $Zn(NO_3)_2$ (Merck) were also used.

2.2 Instrumentation

Microanalytical determinations were carried out in the Cairo University (Microanalytical center). The IR spectra were recorded on a Perkinelmer spectrophotometer (400 – 4000 cm^{-1}) (KBr technique). proton-NMR spectra (DMSO-*d6*) were measured at a Pruker spectrophotometer, using TMS as an internal standard. The uv–vis spectra were recorded on a Perkin-Elmer spectrophotometer. Mass spectra were recorded with the aid of a Shimadzu using a direct insertion probe (DIP) at temperature range 50–800°C. The nano-sized complex was characterized with (SEM) a scanning electron microscope.

2.3 Synthesis of the Schiff base ligand

A solution of 5-bromo-salicylaldehyde (4.02 g, 20 mmol.) in absolute ethanol (50 ml) was added to a hot stirred solution of *o*-phenylenediamine (1.08 g, 10 mmol.). This mixture was refluxed for 3 hr. in water bath and cooling, then, the products formed were collected by filtration. Yield (87.4%, 90.2%); M. P; 240 and 200°C; respectively (the ligand was synthesized in our previous work [24].

2.4 Studied micro-organisms

Isolates of *Pythium ultimum* and *Trichderma harzianum* were obtained from biology department –Jouf University, KSA. Precultures of *Trichderma harzianum* (MW459195) and *Pythium ultimum* (MW830915) were made by grown on potato dextrose agar (PDA) for 7 days at 26°C in dark.

2.5 Preparation of biomass of *Trichderma harzianum* (MW459195)

Two discs (5 mm in diameter) of *T. harzianum* (MW459195) was inoculated onto 100 ml PD broth in a 250 Erlenmeyer flask and incubated at 26°C for 7 days in the dark. Mycelium mats were collected by using filter paper (Whatman No. 1) and washed three times with sterilized distilled water to eliminate any adhering media may present. Mycelium mats (3 g) were dried for bio-synthesis of nanoparticles.

2.6 Pathogenicity tests of nano metal complex against seedling and germination of seeds

In order to use Nano metal complex as bio-control agent, it must not produce metabolic toxins that effect on seed germination and plant growth. Seeds of Fenugreek were soaked in the solution of sodium hypochlorite (2% w/v) for 3 min and washed with sterile distilled H_2O several times. Ten ml of different concentration of nano metal complex (10, 20, 30 and 40 ppm) were prepared and mixed with 0.6 g of carboxymethyl cellulose (CMC). Fifty of Fenugreek seeds were added to the mixture and stirred thoroughly and allowed to dry overnight in a laminar flow cabinet at room temperature, and incubated at 20°C until seeds germinated. Ten ml of sterile distilled H_2O were used as control. After Fenugreek seeds were germinated and beginning to form radicles and plumules, the time of seed germination were determined. Choose the vital seeds and 3 seeds were planted in Erlenmeyer flasks, which containing sterilized 100 ml of WA (3%) was decanted in conical flasks with 10 ml of different concentration of nano-metal complex (10, 20, 30 and 40 ppm). All flasks were incubated in growth chamber at 25°C with 12 h photoperiod (91 μmol m^{-2} s^{-1}) for 4 week. All seeds were then used in study the pathogenicity of nano metal complex on Fenugreek seed germination and seedling elongation.

2.7 Biosynthesis of the nano metal complex

The Zn nano complex was prepared by mixing the Schiff base ligand (1 mmol) and 1 mmol of Zn (II) nitrate in the presence of *T. harzianum* powder as a reducing agent as a novel method for the preparation of Zn-nanocomplex. The mixture was ground with a pestle in an open mortar at room temperature. The melted mixture was then allowed to solidify. The solid was filtered off and crystallized twice-using C_2H_5OH to give Zn-nanocomplex as a pale brown crystal [25].

2.8 Bioassay of nano metal complex against *Pythium ultimum*

Sterilized warm Potato Dextrose Agar (PDA) medium was prepared, and poured in sterile Petri dish supplemented with 100 μl of different concentration of nano metal complex (10, 20, 30 and 40 ppm). Agar discs (10 mm) were taken from the margins colony grown on water agar, and then placed in the center of each Petri dish. All petri-dishes were incubated in the dark at 26°C. Radial fungal mycelial growth was measured by determined colony diameters in each dishes, after 6 days of incubation. The percentage inhibition of fungal growth was calculated using the formula:

$$\text{Inhibition}\,(\%) = \left((\mathrm{D} - \mathrm{M}) \div \mathrm{D}\right) \times 100 \qquad (1)$$

where D is the growth diameter of *P. ultimum* in the control (mm) and M is the growth diameter of *P. ultimum* in presence of Nano-metal complex (ppm) [26].

2.9 Microscopic analysis of the effect of nano metal complex on the mycelium growth

The microscopic images were taken at the faculty of Agriculture, Cairo University, using a scanning electron microscope (JOEL brand model JSM6400) at 1000 and 3000 x. Samples were taken from mycelium treatment with tested nano metal complex at 30 ppm, then the mycelium was placed on a glass slide and using a metal coating equipment of samples (EDWARDS E306A), a copper bath was applied to the samples for 20 min for subsequent visualization. Finally, the general morphological and structural characteristics of the *P. ultimum* mycelium were evaluated.

2.10 Evaluation of nano metal complex as antifungal activity and plant growth promotion in pot experiment

The ability of nano metal complex in different concentrations (10, 20, 30 and 40 ppm), its antifungal and growth enhancement effects of Fenugreek seedlings in soil infested with *P. ultimum* were tested. For preparation *P. ultimum* infested soil, the propagule suspension of *P. ultimum* was added at thirty-propagules/g soil to the autoclaved soil. Thirty seeds were used for each treatment. The surface of the seeds were sterilized with 2% sodium hypochlorite for 3 min, and washed by sterile distilled H_2O for 5 min and dried using sterile filter paper. Pre germination test was done to select viable seeds. The experiment was classified into five groups. Frist group: free *Pythium* soil were irrigated with sterile distilled water, while second groups: soil infested with *Pythium* and irrigated sterile distilled water. The remaining three groups: the soil infested with *Pythium* in each group and soil were irrigated with nano metal complex -solution with different concentrations (10, 20, 30 and 40 ppm). Pots were kept in a growth-illuminated cabinet (Precision, United States) at 25°C with 12 h photoperiod (91 μmol/ m^2 s^1) under humid conditions. Soil moisture content was preserved at 35%. Experiments were performed with ten replicate pots per treatment. Germination % was calculated after 48 and 72 hour by following formula:

$$\text{Germination percentage} = \frac{\text{Number of germinated seeds}}{\text{Total number of seeds}} \times 100 \quad (2)$$

After 10 days of germination, seedlings growth were determined in terms of shoot and root length (mm).

3. Results and discussion

In the present study, the analytical data of the previously prepared Schiff base ligand and its Zn nano metal complex suggest the structures as in **Figure 1**. The Schiff base ligand and its Zn-nano complex were subjected to elemental analyses (C, H, N and O). The results of elemental analyses with molecular formula and the M.P. were in **Table 1**. The results obtained were in well agreement with those of

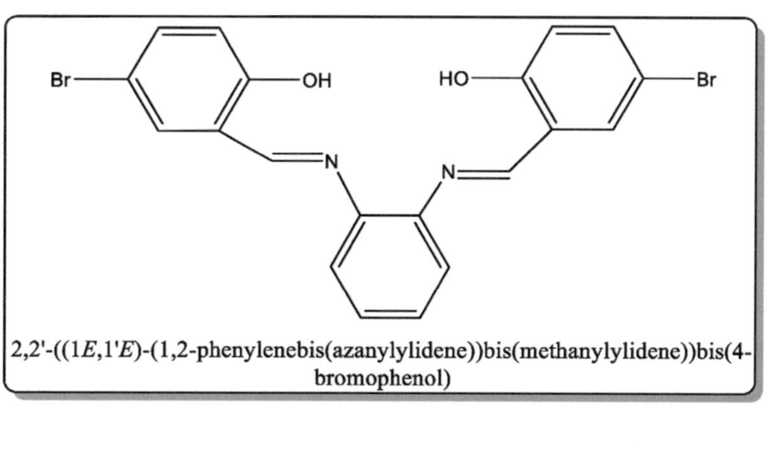

Figure 1.
Schematic route for preparation of the free ligand.

Compd. no. emprical formula	M.P. (°c)	Color (yield %)	(%) found (calcd.) Λm^*			
			C	H	N	
Free ligand(L), $C_{20}H_{14}Br_2N_2O_2$	200	Dark yellow (90.20)	36.89 (50.64)	2.93 (2.98)	12.01 (5.91)	—
(5)[(Zn)(L)]$(NO_3)_2$, $C_{20}H_{14}Br_2N_4O_8Zn$ nano complex	<350	Dark yellow (85.79)	33.59 (33.90)	1.95 (2.00)	7.86 (7.91)	11.25

$^*ohm^{-1}cm^2mol^{-1}$.

Table 1.
Elemental analysis and some physical measurements of the free ligand and its metal nano complex.

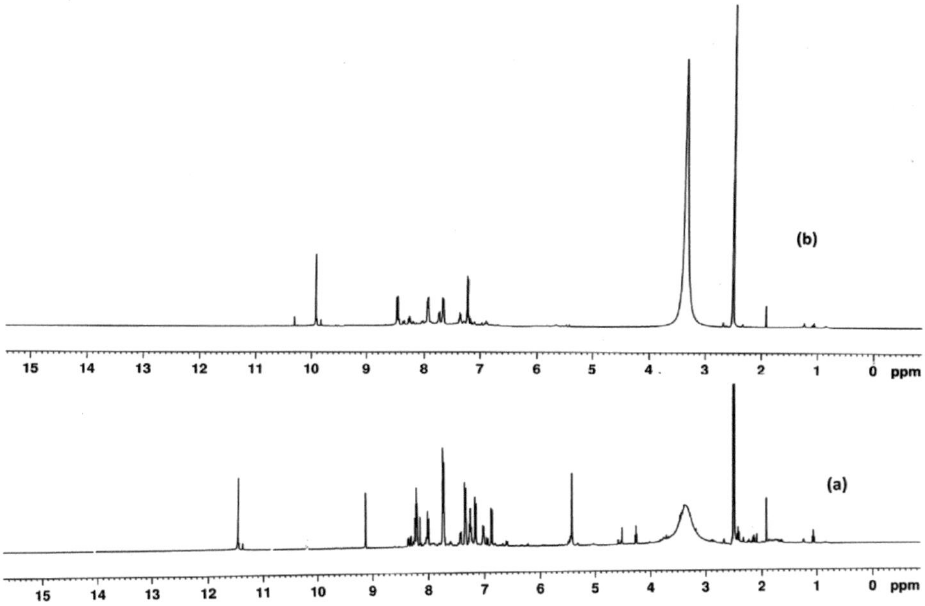

Figure 2.
^1H NMR data for (a) free ligand and (b) its Zn(II) nano complex.

Compd. no.	δ_{ph}	$\delta_{CH=N}$	δ_{OH}
L	6.465–8.175	9.335	11.247
Zn nano complex	6.451–7.684	8.252	10.210

Table 2.
^1H NMR data for the Schiff base ligand and its nano complex.

the suggested formulae. The M.P. were sharp, indicating the purity of the prepared compounds. The results of the elemental analyses of the isolated complex are in agreement with those required by the proposed formula of the complex.

The NMR spectrum of the Schiff base ligand (**Figure 2**) (**Table 2**) showed a singlet peaks at 9.335 ppm and at 11.247 ppm due to the azomethine and phenolic protons, respectively. In addition to multiplet signals at 6.465–8.175 ppm attributed to the aromatic protons [27, 28]. The comparison of the NMR data of the ligand and its Zn (II)metal complex confirms the mode of coordination between the ligand and its metal ion. By complexation, the spectrum of the nano-Zn complex display

Compd. no.	$\upsilon_{O\text{-}H}$ (phenolic)	$\upsilon_{CH=N}$ (azomethine)	$\upsilon_{C\text{-}O}$ (phenolic)	$\upsilon_{M\text{-}O}$	$\upsilon_{M\text{-}N}$	Additional bands
L	3456	1620	1290	—	—	—
Zn-nanocomplex	3401	1532	1240	550	515	1401,1372,1025 (coord.NO_3)

Table 3.
Infra-red spectral data of the free ligand and its metal complex.

a significant shift of the signals due to C=N group and OH protons indicating the involvement of both OH as well as C=N groups in coordination to the metal ions without their deprotonation suggesting that the ligand act as neutral tetradentate ligand, O_2N_2 coordination sphere.

The IR spectrum of the ligand was compared with those of its metal complex, in order to confirm the mode of bonding, **Table 3**.

The infrared spectrum of the schiff base ligand has a broad absorption band at 3456 cm^{-1} which were due to the phenolic group [29, 30]. The spectrum displays also medium band at 1290 cm^{-1} due to $\upsilon_{C\text{-}O}$ (phenolic group) [31]. Additionally, a strong band at1620 cm^{-1} was observed due to $\upsilon_{C=N}$ of azomethine group [32]. By careful comparison of the spectrum of the nano-metal complex with those of the Schiff base ligand it was found that: the band at 3456 cm^{-1} shifted to lower frequency region at 3401 cm^{-1} in the metal nano complex suggesting the involvement of OH group in complexation without its deprotonation. The strong bands at 1620 and 1290 cm^{-1} due to $\upsilon_{C=N}$ and $\upsilon_{C\text{-}O}$ (C=N and OH groups) are shifted to lower wave number (1532 cm^{-1}) and (1240 cm^{-1}), respectively in the metal complex indicating the coordination of nitrogen and oxygen to the metal ion. Also, the spectrum of the complex shows three bands at 1401 cm^{-1}, 1372 cm^{-1} and 1025 cm^{-1} corresponding to unidentate coordination mode of NO_3 group. All these results are consistent with the conductance data. Conclusive evidence of the bonding is also shown by observing new bands in the infrared spectrum of metal nano complex in low frequency region at 550 cm^{-1} and 515 cm^{-1} may be attributed to $\upsilon_{M\text{-}O}$ and $\upsilon_{M\text{-}N}$, respectively that are not observed in the spectrum of the Schiff base ligand [33, 34].

The mass spectrum of the free ligand shows the parent peak at m/e = 474.36 (48.78%) that agree with the molecular formula $C_{20}H_{14}Br_2N_2O_2$. Also, the spectrum shows numerous peaks corresponding to various fragments, their intensity indicates the stability of the fragments. **Figure 3** represent the proposed pathway for the decomposition steps for the ligand.

The conductivity Λ_m value of the Zn-nanocomplex can be calculated using the following relation $\Lambda_m = K/C$, where C is the molar concentration of the nano metal complex solution and K is the specific conductance. The complex was dissolved in (10^{-3} M) DMF and the molar conductivity of the solution at 25 ± 2°C were measured **Table 1**. It is concluded from the results that; the complex is found to has molar conductance value of 11.25 $ohm^{-1}mol^{-1}cm^2$ indicating that this complex is non-electrolytic in nature. Also, the values indicate the bonding of the nitrate ions to metal cations [33, 35].

UV–vis spectra of the ligand and its metal complex was performed in DMF at room temperature at wavelength range 200–800 nm. The significant electronic absorption bands of the ligand, its nano complex and the magnetic moments of the complex are given in **Table 4**. 2 absorption peaks were observed in the spectrum of the ligand at 265 and 339 nm due to $\pi\text{-}\pi^*$ and $n\text{-}\pi^*$ transitions, respectively attributed to benzene and the azomethine groups [34, 36]. In the spectrum of metal complex, the absorption bands attributed to $\pi\text{-}\pi^*$ and $n\text{-}\pi^*$ transition were

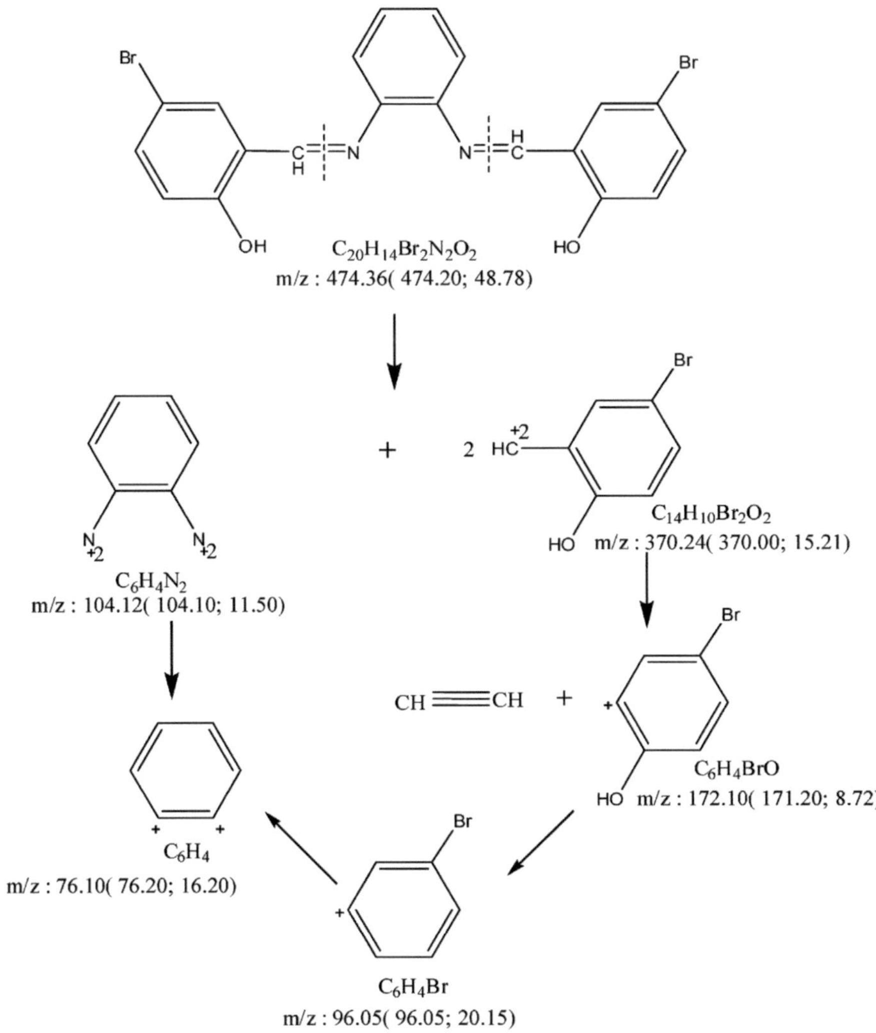

Figure 3.
Mass fragmentation pattern for the free ligand

Compd. no.	μ_{eff} (B.M.)	Absorption bands (nm)		
		π-π*	n-π*	d-d transition
L	—	265	339	—
Zn-nano complex	Diamagnetic	276	323	—

Table 4.
Magnetic moment and electronic spectral data of the ligand and its Zn-nano complex.

shifted to lower or higher frequency due to the coordination of the Schiff base ligand with the metal ions. Zn (II) complex is diamagnetic in nature with no *d-d* transition. So, according to all of this obtained data, the octahedral structure may be suggested.

The size and the morphology of the nano complex were studied by SEM (**Figure 4**). It was found that the particles are semispherical in nature with some agglomerations. The SEM images also revealed the stabilization of Zn(II) nanoparticles due to interaction with the Schiff base ligand [37].

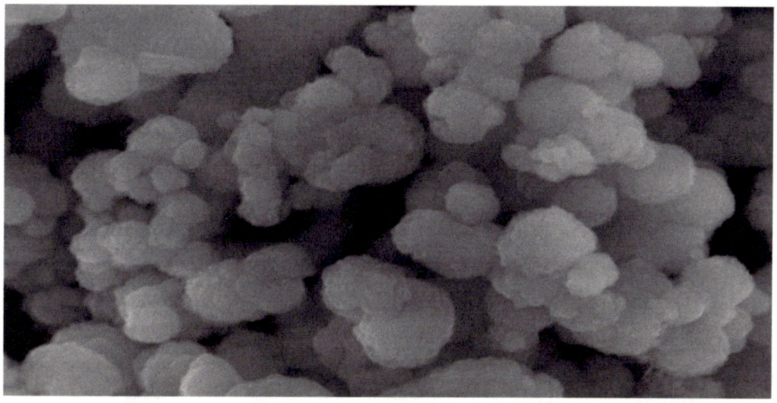

Figure 4.
SEM image of nanoparticle complex.

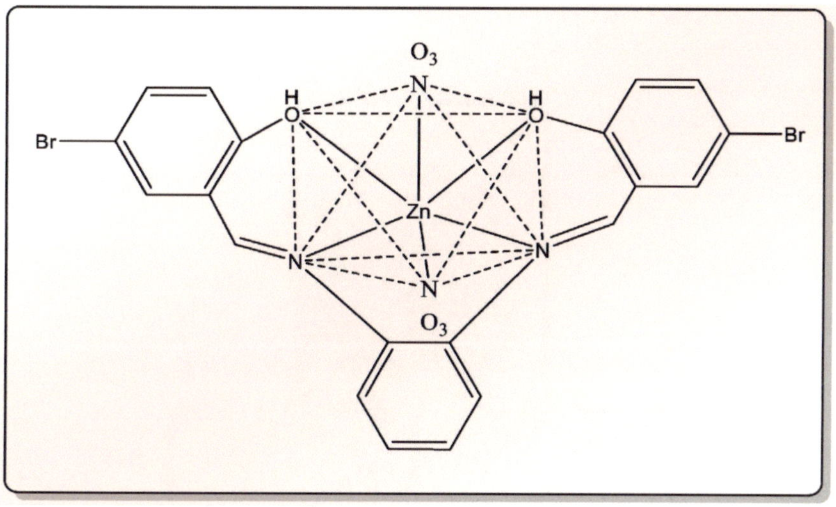

Figure 5.
Suggested structure of Zn nano complex.

Correlation of all this results, the structure of the complex can be suggested to be as in **Figure 5**.

Fungi are the most efficient microorganisms when it comes to the biological production of Mycogenic Metal Nanoparticles synthesis [38]. This is due to the fungi having enhanced processes, accumulating metals, have capacity for producing a high number of bioactive metabolites [39, 40], fungi do not require complex nutrients, these are easy to grow, easy to manipulate, also have high metabolites and production of biomass, and has high metal uptake and high wall-binding capability [41–43]. Our Experiment revealed that nano-metal complex had no obvious effect on the time required for Fenugreek seed germination, (**Figures 6** and 7). Nano-metal complex activated seeds germination, since seeds germinated 24 h prior to the germination of control. Results in **Figure 7** revealed that seedling elongation was retarded with different degrees in case of treated seeds compared with the controls. Nano-metal complex showed a slight retarding effect on Fenugreek seedling elongation compared with the control, especially after 7 days from seed germination.

Table 5 shows the antifungal influence of different concentration of nano metal complex (10, 20, 30 and 40 ppm) on *P. ultimum* mycelial growth on PDA

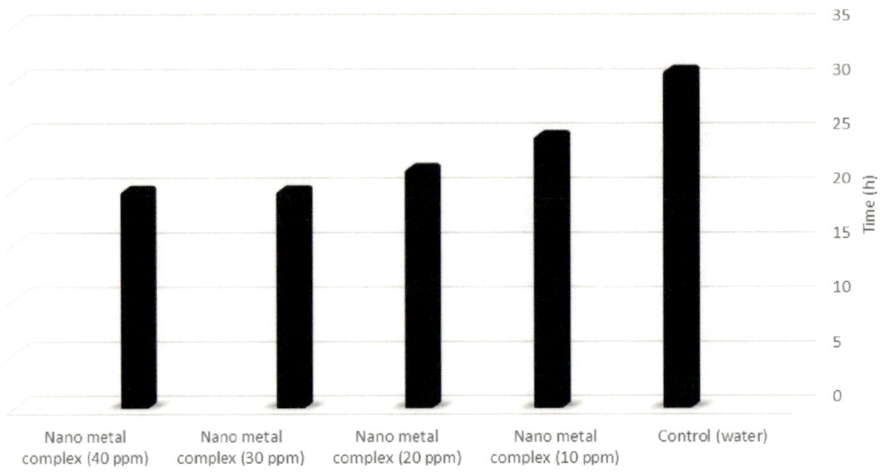

Figure 6.
Effect of nano-metal complex with different concentration on the time of seed germination.

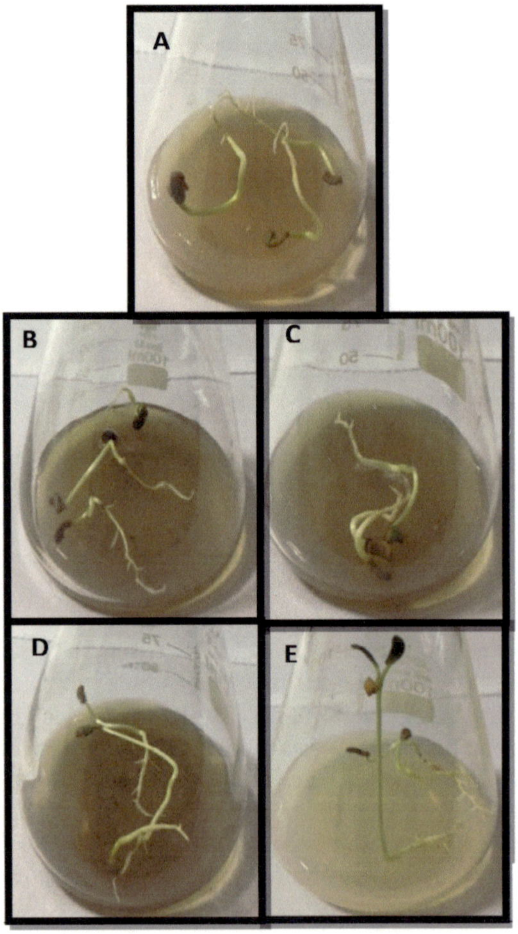

Figure 7.
Pathogenicity tests of nano metal complex against Trigonella foenum-graecum germinating seeds grown in Erlenmeyer flasks containing 2% water agar. (A) Control, (B) treated seeds with nano-metal complex (10ppm), (C) treated seeds with nano-metal complex (20 ppm), (D) treated seeds with nano-metal complex (30 ppm), (E) treated seeds with nano-metal complex (40 ppm) after 7 days at 25°C.

Pythium spp.	Mycelium growth (mm)				
	control (metalaxyl)	Nano metal complex (10 ppm)	Nano metal complex (20 ppm)	Nano metal complex (30 ppm)	Nano metal complex (40 ppm)
P. ultimum	34[**] ±0.16[*]	65 ± 0.3	60 ± 0.13	52 ± 0.21	0

[*]Mycelium growth (mm).
[**]Standard error of three replicates.

Table 5.
Antifungal activity of nano metal complex against P. ultimum *on PDA medium at 26°C for 6 days in the dark.*

medium after 6 days of incubation at 26°C. Results reveled that significantly reduced of mycelial growth of *P. ultimum* at different concentrations (10, 20, 30 and 40 ppm), especially in 30 and 40 ppm compared to control ($p \leq 0.001$). As shown in **Figures 8** and **9** efficacy of nano metal complex at concentration 40 ppm, was the greatest one of all concentration tested *P. ultimum* growth inhibition, that observed no mycelium growth appear, follow by nano metal complex at concentration 30 ppm which mycelium growth were (32 mm). Percentage of inhibition of nano metal complex at different concentrations (10, 20, 30 and 40 ppm) were (23.52%, 29.4%, 62.4% and 100%) respectively. Biological control of damping-off diseases has been successfully. Applied using *T. harzianum* [44, 45]. El-Sayed [46] reported that *Trichoderma* spp. are known to control all pathogens either directly by inhibition of sporulation and growth of the pathogen mechanisms such as enzyme production and mycoparasitism, or indirectly by competing for nutrients and space, modifying the environmental conditions, or by promoting plant growth and enhancing plant defensive mechanisms and antibiosis. Microscopic examination of the mycelium showed coagulation of the protoplasm within the mycelium and lysis of cell wall of mycelium compared with the control (**Figure 10**).

Kamala and Indira [47], Evaluated the activity of three *Trichoderma* isolates (T73, T80 and T105) for their bio-control activity of *P. ultimum*. Also assayed the different bio control mechanisms such as protease, chitinase, β-1,3-glucanase

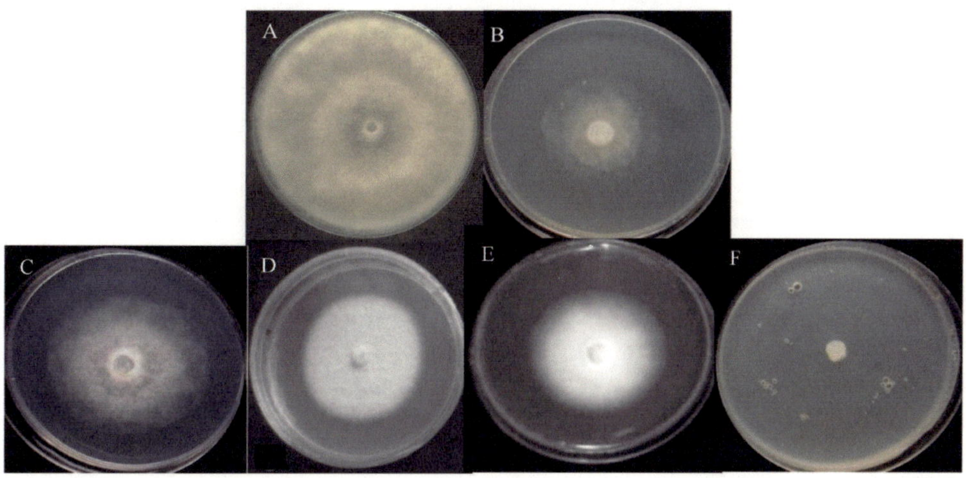

Figure 8.
Inhibition of mycelial growth of P. ultimum. (A) Control, (B) Control containing Metalaxyl, (C) Treated containing Nano-metal complex (10 ppm), (D) Treated containing Nano-metal complex (20ppm), (E) Treated containing Nano-metal complex (30ppm) and F) Treated containing Nano-metal complex (40ppm) on PDA at 26°C for 6 days at the dark.

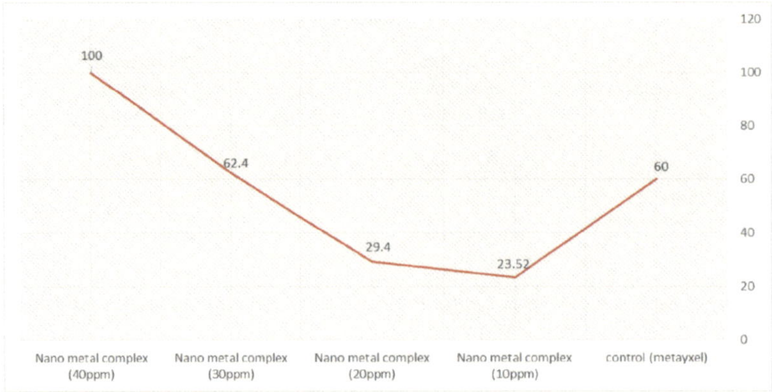

Figure 9.
Inhibition % of P. ultimum exposed to different concentration of nano metal complex, on PDA at 27°C for 6 days at the dark.

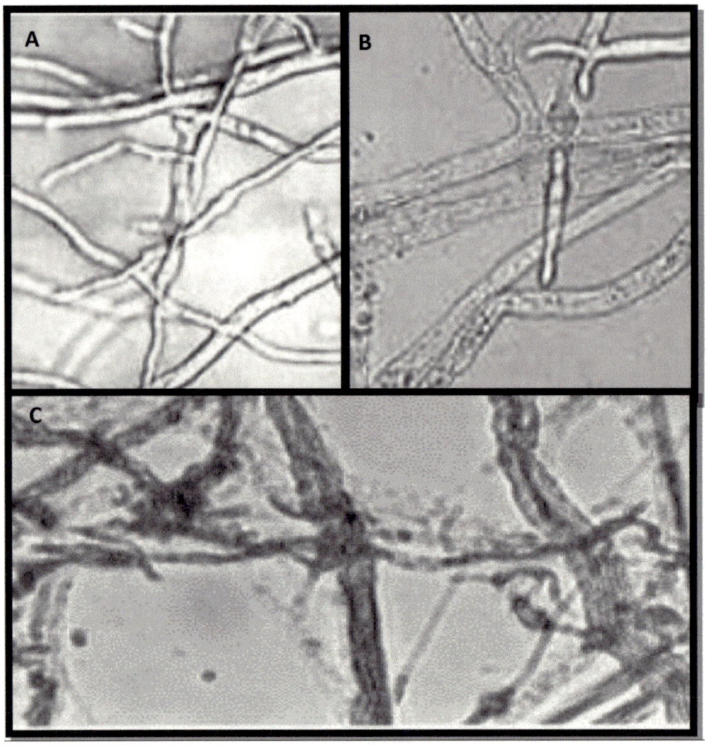

Figure 10.
Scanning electron micrograph showing the antagonistic effect of nano metal complex (30 ppm) against P. ultimum. (A and B) healthy mycelia of P. ultimum. (C) Abnormal structure, lysis and destruction of P. ultimum.

activity, cellulase and production of volatile and non-volatile compounds. Their results show that, *Trichoderma* (T-105) reduced the.

Pre and post-emergence damping-off disease or Root rot disease were incidence in infested soil with *P. ultimum* and showed the highest disease control percentage under in vitro as well as the pot experiment [48–50]. In our study, we evaluated the biogenic Nano-metal complex in pot experiments, which reveal that data in **Table 6** have differences in the germination%, that It observed the beginning of germination

Treatment	Germination % after 20 h	Germination % after 24 h	Germination % after 48 h	Length (mm)	
				Shoot	Root
Control	33.3 ± 0.2	73.33 ± 0.16	100	130	69.5
Nano-metal complex "10 ppm"	60 ± 0.05	93.33 ± 0.13	100	150	81.8
Nano-metal complex "20 ppm"	66.6 ± 0.1	96.67 ± 0.2	100	180	95.2
Nano-metal complex "30 ppm"	66.6 ± 0.12	96.67 ± 0.12	100	180	96.5
Nano-metal complex "40 ppm"	66.6 ± 0.2	96.67 ± 0.22	100	180	96.5

Table 6.
Effect of nano-metal complex treatments on germination (%) and root and shoot lengths (mm) of fenugreek.

of seeds after 20 hour in Nano-metal complex 20, 30 and 40 ppm about 60%,compared with control 33.3%. gradullary, the germination % of seeds increase with increasing time until 48 h, all seeds were germinate 100%.

With observance to root and shoot lengths, results in **Table 6** show that Nano-metal complex had effects in both concentration 20–40 ppm for shoot length about (180 mm), the Nano-metal complex (10 ppm) treatment recorded the lowest value for root length about 150 mm. In addition in the same concentration for root length about (95.2 mm, 96.5 mm and 96.5 mm), respectively. The Nano-metal complex (10 ppm) treatment recorded the lowest value for root length about 81.8 mm, **Figures 11** and **12**. From previous results, it can be concluded that Nano-metal complex (40 ppm) had the best record in both germination as well as shoot and root length.

The increased applications of Metal Nano Particles in several area required new biocompatible, safe, and active nanostructures with less dangerous by products of synthesis reactions [51]. Mycogenic Nano-Particles are observed as biocompatible,

Figure 11.
The effect of bio-control agent and plant growth promotion of nano metal complex on Trigonella foenum-graecum seeds germination in pots containing sandy loam soil cultivated with P. ultimum. (1) Control seeds sown in free Pythium soil, (2–5) seeds sown in soil infested with P. ultimum and irrigated with nano-metal complex (40,30,20, 10 ppm), respectively after 4 weeks.

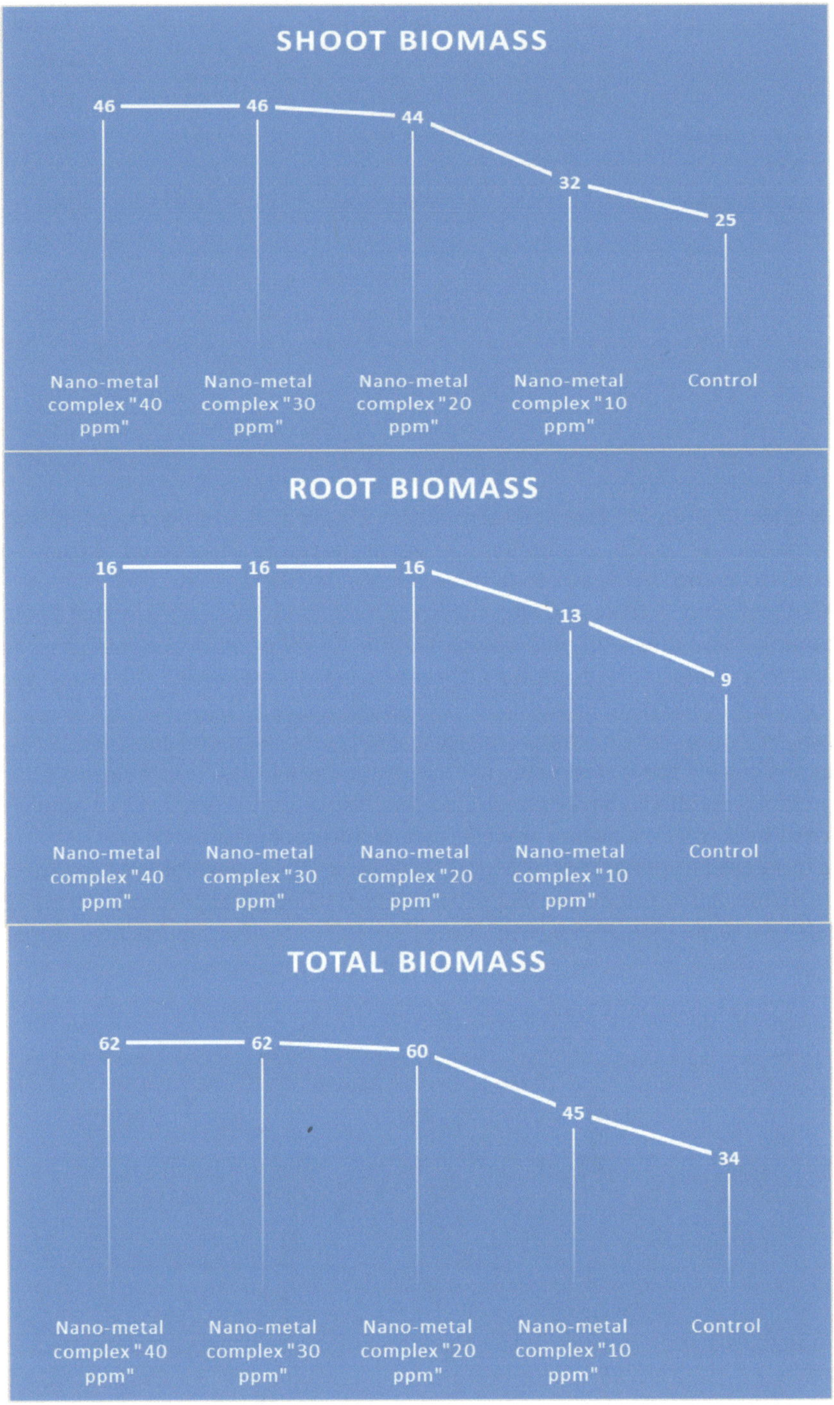

Figure 12.
Effect of nano-metal complex on Trigonella foenum-graecum plant biomass after 4 weeks of planting. (1) Shoot biomass. (2) Root biomass. (3) Total biomass.

eco-friendly, less toxic, safe, and cheapest alternatives, with lowest consumption of energy and highest yields when compared with other physical or chemical synthesis [26, 52]. While well-known advantages of Metal Nano-Particles as the best alternatives against antimicrobial species of pathogenic fungi and other microflora, many challenges and opportunities are ahead of us. Given that the effects of NPs result from a combination of multiple, the potential development of resistance against them is more difficulty and less likely [53]. In future should be focus on testing new isolates, discovering new Metal Nano-Particles, and clarifying their structures and mode of its action as antimicrobial agents. Novel discovery of mycogenic MNPs include the study of extremophilic and endophytic isolates. While the endophytic has increased its relevance during the last years, the extremophilic is still limited in studies and focused on other purpose rather than their use against antimicrobial species [54].

Acknowledgements

Authors are grateful to the College of science, Jouf university-Saudi Arabia for providing the facilities to carry out this work.

Funding information

No funding this work.

Conflict of interest

No conflict of interest.

Institutional review board statement

Not applicable.

Informed consent statement

Not applicable.

Data availability statement

The data presented in this study are available upon request from the corresponding author.

Author details

Shaima M.N. Moustafa[1,3*] and Rania H. Taha[2,4]

1 Biology Department, College of Science, Jouf University, Sakaka, Saudi Arabia

2 Chemistry Department, College of Science, Jouf University, Sakaka, Saudi Arabia

3 Department of Botany and Microbiology, Faculty of Science, Minia University, El-Minia, Egypt

4 Department of Chemistry, Faculty of Science (Girls), Al-Azhar University, Cairo, Egypt

*Address all correspondence to: shymaa.nabil@ju.edu.sa

References

[1] Sánchez-Montesinos B, Diánez F, Moreno-Gavira A, Gea FJ, Santos M. Plant Growth Promotion and Biocontrol of *Pythium ultimum* by Saline Tolerant *Trichoderma* Isolates under Salinity Stress. Int J Environ Res Public Health. 2019 Jun;**16**(11):2053

[2] FRAC, 2016 [internet]. Fungicide Resistance Action Committee. Resistance tables Benz. [cited 2017 Aug 17]. Available from: http://www.frac.info/docs/default-source/expert-fora/2004-06-11resistance_tables_benz.pdf.

[3] Guilger-Casagrande M, de Lima R. Synthesis of silver nanoparticles mediated by fungi: A review. Front. Bioeng. Biotechnol.

[4] Vijayakumar M, Priya K, Nancy FT, et al. Biosynthesis, characterization and anti-bacterial effect of plant-mediated silver nanoparticles using Artemisia nilagirica. Ind Crops Products. 2013;41:235_240.

[5] Peter L, Silambarasan S, Abraham J. Ecofriendly synthesis of silver nanoparticles from commercially available plant powders and their antibacterial properties. Scientia Iranica. 2013;20:1049_1054.

[6] Peter L, Sivagnanam S, Abraham J. Synthesis of silver nanoparticles using plants extract and analysis of their antimicrobial property. J Saudi Chem Soc. 2015; 19:311_317.

[7] Gade A, Ingle A, Bawaskar M, et al. Fusarium solani: a novel biological agent for the extracellular synthesis of silver nanoparticles. J Nanoparticle Res. 2009;11:2079_2085.

[8] Neveen-Mohamed K. Biogenic silver nanoparticles by Aspergillus terreus as a powerful nanoweapon against Aspergillus fumigates. Afr J Microbiol Res. 2014;7: 5645_5651.

[9] Husseiny MI, Aziz MAE, Badr Y, et al. Biosynthesis of gold nanoparticles using *Pseudomonas aeruginosa*. Spectrochimica Acta Part A. 2006;67:1003_1006.

[10] El-Shanshoury AER, ElSilk SE, Ebeid ME. Extracellular biosynthesis of silver nanoparticles using *Escherichia coli* ATCC 8739, *Bacillus subtilis* ATCC 6633, and *Streptococcus thermophilus* ESh1 and their antimicrobial activities. ISRN Nanotechnol. 2011;**2011**:385480

[11] Dueke-Eze, C. U., T. M. Fasina and N.Idika,2011, African Journal of Pure and Applied Chemistry, 5(2) 13-18.

[12] Ali E. S¸abik, ork MuharremKarab, G¨okhan Ceyhan, Mehmet T¨umer and Metin Dı˘grak, 2012, International Journal of Inorganic Chemistry, (2012)1-11.

[13] Jain SC, Agrawal M, Sharma RA. The genus Trigonella—phytochemistry and biology. Anc Sci Life. 1996;**16**:108-117

[14] Basch E, Ulbricht C, Kuo G, et al. Therapeutic applications of fenugreek. Altern Med Rev. 2003;**8**:20-27 (Review)

[15] Roberts KT. The potential of fenugreek (*Trigonella foenum-graecum*) as a functional food and nutraceutical and its effects on glycemia and lipidemia. J Med Food. 2011;**14**:1485-1489

[16] Weder JK, Haussner K. Inhibitors of human and bovine trypsin and chymotrypsin in fenugreek (*Trigonella foenum-graecum* L.) seeds. Isolation and characterization. Zeitschrift fur Lebensmittel-Untersuchung undForschung. 1991;**192**:535-540

[17] Weder JK, Haussner K. Inhibitors of human and bovine trypsin and

chymotrypsin in fenugreek (*Trigonella foenum-graecum* L.) seeds.
Reaction with the human and bovine proteinases. Zeitschrift fur LebensmittelUntersuchung und-Forschung. 1991;**193**:321-325

[18] Abdo MS. al-Kafawi AA. Experimental studies on the effect of Trigonella foenum-graecum. Planta Med. 1969;**17**:14-18

[19] Hyder, S.; Inam-ul-Haq, M.; Ashfaq, M.; Ahmad, A.; Gondal, A.S.; Iqbal, M. First report of Pythium myriotylum D., causing damping off and root rot in chili pepper (Capsicum annum L.) from Punjab, Pakistan. Plant Dis. 2018, 102, 687.

[20] Ali M, Shahid AA, Subhani MN. Maping and monitoring for the valuation of soil fungi and chili damping off. J. Anim. Plant Sci. 2019;**29**:737-745

[21] Hyder, S.; Inam-ul-Haq, M.; Ashfaq, M.; Ahmad, A.; Gondal, A.S.; Iqbal, M. First report of Pythium myriotylum D., causing damping off and root rot in chili pepper (Capsicum annum L.) from Punjab, Pakistan. Plant Dis. 2018, 102, 687

[22] Ali M, Shahid AA, Subhani MN. Maping and monitoring for the valuation of soil fungi and chili damping off. J. Anim. Plant Sci. 2019;**29**:737-745

[23] Nawaz K, Shahid AA, Subhani MN, Anwar W, Aslam M. First Report of Pythium spinosum Causing Root Rot of Chili (Capsicum annuum) in Pakistan. Plant Dis. 2016;**100**:526

[24] Rania H Taha, Preparation, Spectroscopic Characterization, DNA Cleavage, Antimicrobial and Antitumor Investigations of Nickel and Uranyl Schiff Base Complexes in Bulk and Nano size, Current Science International, 4(4), 2015

[25] Devi TP, Kulanthaivel S, Kamil D, Borah JL, Prabhakaran N, Srinivasa N. Biosynthesis of silver nanoparticles from *Trichoderma* species. Indian J. Exp. Biol. 2013;**51**:543-547

[26] Guilger-Casagrande M, Germano-Costa T, Pasquoto-Stigliani T, Fraceto LF, de Lima R. Biosynthesis of silver nanoparticles employing *Trichoderma harzianum* with enzymatic stimulation for the control of Sclerotinia sclerotiorum. Sci. Rep. 2019;**9**:1-9

[27] Fasina TM, Ogundele OO, Ayeni I. Journal of. Chemical and Pharmaceutical Research. 2014;**6**(6):816-819

[28] Gomathi,V. and R. Selvameena, 2013, INDIAN JOURNAL OF APPLIED RESEARCH, 3(4), 51-53.

[29] Aliyu HN, Bello I. Biokemistri. 2010;**22**(2):105-110

[30] EL-Hashash, M. A., Essawy, A. and Fawzy, A. S., 2014, Advances in Chemistry, (2014), 1-10.

[31] Kumar M, Saxena PN. ORIENTAL. JOURNAL OF CHEMISTRY. 2012;**28**(4):1927-1931

[32] Mahendra Raj Karekal, Vivekanand Biradar and Mruthyunjayaswamy Bennikallu Hire Mathada, 2013, Bioinorganic Chemistry and Applications, (2013), 1-16.

[33] Khandelwal, A., M. Agrawal, S. Lamba and G. Baswal, 2013, Weekly Science Research Journal, (1), 1-10.

[34] Riyadh M. Ahmed, Enaam I. Yousif, Hasan A. Hasan, and Mohamad J. Al-Jeboori, 2013, The Scientific World Journal, (2013), 1-7.

[35] Mohamed,G.G.; Zayed,M.A. and Abdallah,S.M., J. Mol. Struct., (2010), 979, 65.

[36] Mohebbi S, Bakhshi B. J. Coord. Chem. 2008;**61**:2615

[37] Mandegani, Z., A. Mozaffar, A. Zahra, M. Afshan, I. Nasser and O.

Akbar, 2015, Green Chem.,17, 3326-3337.

[38] Khan AU, Malik N, Khan M, Cho MH, Khan MM. Fungi-assisted silver nanoparticle synthesis and their applications. Bioprocess Biosyst. Eng. 2017;**41**:1-20

[39] Simões MF, Ottoni CA, Antunes A. Biogenic metal nanoparticles: A new approach to detect life on Mars? Life. 2020;**10**:28

[40] Ovais M, Khalil AT, Ayaz M, Ahmad I, Nethi SK, Mukherjee S. Biosynthesis of metal nanoparticles via microbial enzymes: A mechanistic approach. Int. J. Mol. Sci. 2018;**19**:4100

[41] Zhao X, Zhou L, Rajoka MSR, Yan L, Jiang C, Shao D, et al. Fungal silver nanoparticles: Synthesis, application and challenges. Crit. Rev. Biotechnol. 2018;**38**:817-835

[42] Khandel P, Shahi SK. Mycogenic nanoparticles and their bio-prospective applications: Current status and future challenges. J. Nanostruct. Chem. 2018;**8**:369-391

[43] Yadav A., Verma A., Yadav K. *Advance Appliance through Fungal Nanobiotechnol.* Springer; Cham, Switzerland: 2016. Fungal nanoparticles: An emerging tool in medical biology; pp. 213-240.

[44] Ragab MMM, Abada KA, Abd-El-Moneim LM, Abo-Shosha YZ. Effect of different mixtures of some bioagents and Rhizobium phaseoli on bean damping-off under field condition. Inter. J. Sci. Eng. Res. 2015;**6**(7):1099-1106

[45] Ahmed GA. Efficiency of some antioxidants and bioagents in controlling Rhizoctonia damping-off of snap bean. Middle East J. App. Sci. 2016;**6**(4):748-758

[46] Sahar A. El-Sayed, Management of Damping-Off and RootRot Diseases of Faba bean by Bioproducts and Inducer Resistance Chemicals. Egypt. J. Phytopathol., Vol. 45, No. 1, pp. 135-156 (2017)

[47] Kamala T, Indira S. Evaluation of indigenous *Trichoderma* isolates from Manipur as biocontrol agent against *Pythium aphanidermatum* on common beans. 3. Biotech. 2011;**1**:217-225

[48] El-kafrawy AA. Biological control of bean damping-off caused by *Rhizoctonia solani*. Egypt J Agric Res. 2000;**80**(1):57-70

[49] Gonzalez RM, Castellanos LG, Rames MF, Perez GG. Effectiveness of *Trichoderma* spp. for the control of seed and soil pathogenic fungi in bean crop. Fitosanidad. 2005;**9**(1):37-41

[50] Malik G, Dawar S, Sattar A, Dawar V. Efficacy of *Trichoderma harzianum* after multiplication on different substrates in the control of root rot fungi. Int J Biol Biotechnol. 2005;**2**(91):237-242

[51] Yadav A, Kon K, Kratosova G, Duran N, Ingle AP, Rai M. Fungi as an e_cient mycosystem for the synthesis of metal nanoparticles: Progress and key aspects of research. Biotechnol. Lett. **2015**;**37**:2099-2120

[52] Singh R, Nawale LU, Arkile M, Shedbalkar UU, Wadhwani SA, Sarkar D, et al. Chemical and biological metal nanoparticles as antimycobacterial agents: Acomparative study. Int. J. Antimicrob. Agents. **2015**;**46**:183-188

[53] Minakshi P, Ghosh M, Brar B, Kumar R, Lambe UP, Ranjan K, et al. Nano-antimicrobials: A new paradigm for combating mycobacterial resistance. Curr. Pharm. Des. **2019**;**25**:1554-1579

[54] Mourato A, Gadanho M, Lino AR, Tenreiro R. Biosynthesis of crystalline silver and gold nanoparticles by extremophilic yeasts. Bioinorg. Chem. Appl. **2011**;**2011**:546074

Chapter 8

Organic Fertilization with Residues of Cupuassu (*Theobroma grandiflorum*) and Inga (*Inga edulis*) for Improving Soil Fertility in Central Amazonia

Eleano Rodrigues da Silva, Marta Iria da Costa Ayres, Acácia Lima Neves, Katell Uguen, Luiz Antonio de Oliveira and Sonia Sena Alfaia

Abstract

The cupuassu (*Theobroma grandiflorum* (Willd. Ex Spreng.) K. Schum.) is a native fruit tree which has, in the past years, acquired great social and economic importance for the regional farmers. The nutrient-rich and often wasted cupuassu tree fruit shell residues can contribute to the improvement of the low fertility soil of Amazonia. A trial was carried out on a small holder's cupuassu plantation in Central Amazonia to ascertain the effect of organic fertilization on the recovery of soil fertility and plant nutrition by using material from cupuassu shell residues and *Inga edulis* pruning (branches and leaves). The fertilization with cupuassu rinds + Inga prunings improved soil fertility, mainly by the increase of K and Ca in the soil, but only with liming, which appears to favor the mineralization of these nutrients. At the 0–10 cm depth, the Ca level increased about 50% compared to the control and the K level increased 75% compared to the cupuassu shell treatments. The significant increase of about 30% in N absorption by trees in the plots without liming shows that the application of green manure can increase the mineralization of N in Oxisols. These results show that the organic residue sources used can result in a nutrient-bearing organic fertilizer and become a low-cost alternative for recycling cupuassu processing residues.

Keywords: cupuassu shell, Inga pruning, decomposition, nutrients, liming

1. Introduction

The cupuassu (*Theobroma grandiflorum* (Willd. Ex Spreng.) K. Schum.) is a native fruit tree which has, in the past years, acquired great social and economic importance to the regional farmers. Most soils, where this species is cultivated, are of low natural fertility. Successive crops with no nutrient replenishment can bring about soil exhaustion and become detrimental to the system's sustainability [1, 2]. Problems, such as population unevenness, nutritional deficiencies, the occurrence

Of witch's broom disease caused by a fungi (*Moniliophtora perniciosa*), and cultivation of plants susceptible to it, have imposed a crop yield decrease, causing losses to the rural producers [2–4].

The municipality of Presidente Figueiredo in the Amazonas state has great potential for cupuassu agribusiness, despite the producers have been faced with problems related to the decrease in cupuassu yield due to the soil's natural fertility loss throughout years of cultivation, plus the occurrence of pests and diseases. Since most producers in this municipality cannot afford agricultural inputs, an alternative technology for conserving and recovering soils must be economically viable to be adopted by the farmers in the region. Hence, low-cost soil management alternatives, such as the use of organic residues, must be prioritized.

The often wasted cupuassu tree fruit shells could be used in contributing to fertilize the soils. The recycling of these wastes, as well as a better use of natural resources, has become very important when it comes to sustainable agriculture since it reduces production costs and minimizes environmental impact [5, 6]. Furthermore, organic residues can bear phytopathogenic suppressing microorganisms present in the soil, in addition to stimulating the growth of a more diversified and competitive/suppressing microbiota, reducing the occurrence of plant diseases [7, 8].

As most fruits, especially cupussu, are rich in potassium (K) and their productivity depends on the contents of this nutrient in the soil, their replenishment becomes necessary to maintain the cupuassu productivity [1, 9, 10]. Large amounts of organic residues from cupuassu pulping processing are currently found available piled up near the dwellings and/or agroindustries in the municipality of Presidente Figueiredo. These residues are a low-cost fertilizer alternative source for the replenishment of nutrients exported by harvesting or lost by leaching. On the other hand, the mixing of the cupuassu shells with N-rich materials could help the cupuassu shells' decomposition and mineralization without the soil's N immobilization in the soil. Thus, the addition of pruning of some N_2-fixing leguminous species, such as the Inga tree (*Inga edulis* Mart.), also known as ice cream bean, which is very frequent in the properties of the farmers of the Amazonas, might be used for this purpose. Hence, an experimental assay was implemented in the site of a smallholder farmer in the municipality of Presidente Figueiredo, in Central Amazonia, aiming to evaluate the effect of an organic fertilizer, by using plant materials from cupuassu shell residues and Inga tree pruning, on recovering the fertility of an Oxisol.

2. Material and methods

2.1 Characterization study area and experimental design

This study was conducted on a farming site located at coordinates 02° 03′ 55.5″ S and 59° 22′ 55.5″ W, on the side roads called Morena, in the municipality of Presidente Figueiredo, about 150 km from Manaus-AM through highway BR-174, in the Central Amazonia region. The experimental site's soil was classified as Oxisol, with flat to slightly undulated relief and belonging to the very clayey textural class, with the following chemical characteristics (0–20 cm layer): pH (H_2O) 4.62; available P 3.67 mg kg^{-1}; total N 1.88 g kg^{-1}; and 0.36; 0.16; 0,07; 2.75 cmo_c kg^{-1} of exchangeable Ca, Mg, K, and Al, respectively.

The experiment was carried out in an eight-year-old cupuassu crop, with 5 × 5 m spacing between plants and 400 plants ha^{-1} density. The experimental design was a randomized block design with four blocks (replications), consisting of a 2 × 3 factorial scheme, in which the factors were two levels of liming and three types of fertilization, making a total of six treatments, totaling 24 plots with 10 cupuassu plants per plot and a total of 60 plants per block. Treatments were as follows: 1—soil control; 2—cupuassu shell 2 t ha^{-1}; 3—cupuassu shell 2 t ha^{-1} + Inga prunings 3 t ha^{-1}; 4—Limestone 2 t ha^{-1}; 5—cupuassu shell 2 t ha^{-1} + limestone 2 t ha^{-1}; 6—cupuassu shell 2 t ha^{-1} + Inga prunings 3 t ha^{-1} + limestone 2 t ha^{-1}.

The dolomitic limestone (2 t ha^{-1}) was applied within the area straight below the cupuassu plant canopy and superficially incorporated into the soil. Plant materials used as organic fertilizers were obtained from farmers growing sites. Cupuassu shells were removed from a pile of discarded cupuassu shells and ground in a disintegrator (chopper and grinder), whereas the plant material was obtained from adult Inga tree pruning. Pruning was done by cutting off branches bearing up to 1.0-cm-thick leaves, just prior to the fertilization. Organic fertilizers were applied within the area straight below the cupuassu tree canopies, 60 days after the limestone was applied.

Plant materials used as organic fertilizer were analyzed to determine their chemical characteristics according to Silva [11] methodology. Inga plant material held higher nutrient concentrations than those of cupuassu shell, except for K, the contents of which showed to be similar (**Table 1**).

2.2 Soil sampling and analyses

Soil and cupuassu leave samplings were carried out at the end of the 2007 harvest, so as to evaluate soil fertility and plant nutritional status. Soil samples, within the area below plant canopies, were collected at 0–10 and 10–20-cm deep, for a total of 10 simple subsamples to make up a composite sample, and 20–30-cm deep, in a total of 5 subsamples to form one composite sample. Soil samples were air-dried, sieved through a 2-mm mesh, and taken to INPA's Soil and Plants Laboratory, where they were analyzed according to the methodology used by Embrapa [12]. Soil pH was measured in water at a ratio of 1:2.5. The cations of Ca, Mg, and Al were extracted using KCl 1 N, and their concentration was measured using atomic absorption spectrophotometry. For P and K, the double acid extraction system (H_2SO_4 0.0125 M + HCl 0.05 M) was used. P levels were determined by spectrophotometry using ammonium molybdate. The organic carbon concentrations were obtained using the self-analyzer for C, H, and N from Carlo Erba manufacturer.

Four simple soil samples were collected at 0–5, 5–10, and 10–20-cm deep, within the area straight below the canopies of the four central plants in the plot, for 0, 30, 60, and 90 days following the organic fertilizers application, so as to determine the N in the ammonium (NH_4^+) and nitrate (NO_3^-) forms. Samples were placed in plastic bags and stored in a styrofoam box containing ice. They were then transported to the laboratory and weighed for mineral-N extraction. Mineral N contents were determined after having been extracted with 0.5 M K_2SO_4, by using 20 g of moist soil and 40 mL of K_2SO_4. Nitrate and ammonium contents were determined through colorimetry following Embrapa [12] methodologies.

The sixth recently ripened sprouting leaf, downward from the tip of one of the mid-canopy branches, was established as a standard reference in the collection of

Plant material	N	Ca	Mg	K	P	C	C/N
			g kg^{-1}				
Cupuassu shell	6.19	1.29	0.90	6.58	0.33	495	68.3
Pruning of Inga	24.90	6.99	1.65	6.49	1.15	483	19.4

Table 1.
Nutrient concentrations in plant materials used as organic fertilizer, collected on farms in the Central Amazon region (n = 3).

cupuassu leaves. A total of five leaves were removed from each useful plant per plot, rapidly cleaned with distilled water-soaked cotton, dried in a forced ventilation oven at 65°C for three days, and then ground. Ca, Mg, K, and P concentrations were determined by nitro-perchloric digestion and the total N through sulfur digestion followed by distillation through the micro-Kjeldahl method [12].

2.3 Fruit production

The fruit production was evaluated by the number and weight of cupuaçu of 10 useful plants of each plot, from December to April (time of production cultivation), during the 2007/2008 harvest. The fruits, once detached from the plant, were counted and weighed. This operation was carried out once a week, during the entire harvest period.

2.4 Statistical analysis

Treatment effects significance was determined by the analysis of variance (ANOVA), and the comparisons between the means of the variables were performed by the Tukey test at 5% probability.

3. Results and discussions

3.1 Soil acidity and calcium and magnesium contents

As it was expected, liming significantly increased soil pH, Ca, and Mg contents and reduced the exchangeable Al at the three assessed depths (**Table 2**). Despite the 2 t ha^{-1} of limestone application significantly reduced the exchangeable Al content at the three evaluated depths, it still remained high at deeper layers (1.01–2.00 cmo_c kg^{-1}) [13]. The liming effect was more pronounced in the 0–10 cm layer, mainly in the treatments with the application of organic residues. There occurred an increase of the soil pH, as well as a more pronounced reduction of the Al content in the treatments with an application of organic residues, in the limestone plots, especially in the superficial layer of the soil. The increase in soil pH with the addition of plant residues has been observed [14]. Castro [15] observed Al content marked reduction with the application of organic fertilizers on an Ultisol in Central Amazonia. The intensity of the effects is linked to the characteristics of the plant material used. In general, legumes provide higher pH and Al neutralization in the soil than grasses, and this effect is linked to the cation content in the plant material [16]. This result would be due to the complexation of free H^+ and Al^{3+} with anionic organic compounds from the residues and the increased saturation of the cation exchange capacity by Ca, Mg, and K added with the plant residue, which would reduce the potential acidity [14]. The Oxisols of the Amazon are characterized by high acidity and the presence of toxic Al, and the application of organic matter has been suggested as an alternative for correcting the acidity and the neutralization of the exchangeable Al in the soil [17, 18].

3.2 Exchangeable K, available P, and organic C contents

The results of the **Table 3** showed a beneficial effect of liming on the decomposition of the organic material added, resulting in the increase of the mineralization of K in the soil, improving the efficiency of the organic fertilizers used. In that case, K concentration on the plant materials used on the organic fertilization may

Treatments	Soil depth (cm)					
	0–10		10–20		20–30	
	Without lime	With lime	Without lime	With lime	Without lime	With lime
			pH (H_2O)			
Controls	4.29	4.87 b	4.25	4.64	4.30	4.43
Cupuassu shell[1]	4.30	5.08 ab	4.25	4.66	4.26	4.51
Cupuassu shell + Inga2	4.24	5.31 a	4.20	4.60	4.20	4.39
Mean lime	4.28 B	5.09 A	4.23 B	4.63 A	4.25 B	4.44 A
			Al ($cmol_c\ kg^{-1}$)			
Controls	2.98	1.01	2.96	1.95	2.64 b	2.36
Cupuassu shell[1]	2.88	0.53	2.64	1.66	2.52 b	2.21
Cupuassu shell + Inga2	3.13	0.34	3.03	1.75	3.27 a	2.31
Mean lime	3.00 A	0.63 B	2.88 A	1.79 B	2.81 A	2.29 B
			Ca ($cmol_c\ kg^{-1}$)			
Controls	0.65	2.02 b	0.28	0.90	0.12	0.42
Cupuassu shell[1]	0.51	2.45 ab	0.23	0.95	0.10	0.45
Cupuassu shell + Inga2	0.32	3.02 a	0.22	0.86	0.12	0.35
Mean lime	0.49 B	2.50 A	0.24 B	0.90 A	0.11 B	0.41 A
			Mg^{2+} ($cmol_c\ kg^{-1}$)			
Controls	0.37	1.03	0.17	0.57	0.10	0.30
Cupuassu shell[1]	0.32	1.10	0.16	0.57	0.09	0.33
Cupuassu shell + Inga[2]	0.28	1.19	0.14	0.60	0.09	0.27
Mean lime	0.32 B	1.17 A	0.16 B	0.58 A	0.09 B	0.30 A

Means followed by different lowercase letters in the columns, and different upper case letters in the lines, differ from each other, by Tukey test at 5% probability.
12 t ha^{-1} organic fertilization of residues of cupuassu shell.
22 t ha^{-1} organic fertilization of residues of cupuassu shell +3 t ha^{-1} Inga pruning.

Table 2.
Values of pH (H_2O), exchangeable aluminum (Al), calcium (Ca), and magnesium (Mg) in an Oxisol, cultivated with cupuassu tree, as a function of lime and organic fertilizer.

have been the determinant factor for such an effect, since in the treatment with *cupuassu shells + Inga prunings*, we applied almost twice the amount of K we had done in the treatment with just cupuassu shells. Similar results have also been observed by Alfaia et al. [9] in cupuassu agroforestry systems in Western Amazonia.

In general, the average levels of K in the soil observed in this study were low (<0.05–0.10 $cmol_c\ kg^{-1}$), according to Moreira and Fageria [13] criteria. Soils in Amazonia hold low K contents [6, 13, 19, 20] and, due to the relatively high demand many native plants have for it, K becomes one of the most limiting nutrients for

Treatments	Soil depth (cm)					
	0–10		10–20		20–30	
	Without lime	With lime	Without lime	With lime	Without lime	With lime
			K ($cmol_c$ kg^{-1})			
Controls	0.09 A	0.09 Aab	0.05 A	0.05 Aab	0.04	0.03
Cupuassu shell[1]	0.12 A	0.07 Ab	0.06 A	0.04 Bb	0.04	0.03
Cupuassu shell + Inga[2]	0.09 B	0.15 Aa	0.04 B	0.07 Aa	0.04	0.04
Mean lime	0.10	0.10	0.05	0.05	0.04	0.03
			P (mg kg^{-1})			
Controls	4.4	4.6	2.5	3.6	1.7	2.9
Cupuassu shell[1]	4.6	3.9	2.8	2.8	1.8	2.1
Cupuassu shell + Inga[2]	4.3	3.5	2.8	2.5	2.0	2.0
Mean lime	4.4	4.0	2.7	3.0	1.8 B	2.3 A
			Organic C (g kg^{-1})			
Controls	35.0	28.4	24.2	21.6	19.4	19.1
Cupuassu shell[1]	32.9	29.4	22.3	20.5	17.4	17.3
Cupuassu shell + Inga[2]	31.1	29.1	21.7	19.8	18.3	17.1
Mean lime	33.0 A	28.9 B	22.7 A	20.6 B	18.4	17.8

Means followed by different lowercase letters in the columns, and different upper case letters in the lines, differ from each other, by Tukey test at 5% probability.[1]*2 t ha^{-1} organic fertilization of residues of cupuassu shell.*
[2]*2 t ha^{-1} organic fertilization of residues of cupuassu shell +3 t ha^{-1} Inga pruning.*

Table 3.
Values of exchangeable potassium (K), available phosphorus (P), and organic carbon (C) in an Oxisol, cultivated with cupuassu tree, as a function of lime and organic fertilizer.

producing fruits in this region [21]. Studies conducted in the Manaus region have shown K to be the most exported nutrient through agroforestry products originating from Amazonian native species, such as cupuassu, peach palm (*Bactris gasipaes*), and assai (*Euterpe oleracea*), with higher concentration in shells, seeds, and infructescence petioles, which must be reincorporated into the planting areas to maintain sustainability [22].

The low exchangeable K contents in the soils may also be related to its export through fruit harvesting in addition to losses by leaching, according to what was observed by Alfaia et al. [9]. The results of that study confirmed that mineralization and, the addition to the soil, of K-rich organic matter, such as cupuassu shells and Inga plant material, have the potential to restore this nutrient in the soil when associated with the correction of the soil acidity [9, 23].

With regard to P, it was observed that only at a depth of 20–30 cm the levels of this nutrient were significantly higher in the presence of liming, although the data in **Table 4** show a significant effect of liming on the absorption of P by cupuassu plants.

According to Khorramdel et al. [24], soil organic matter content is a result of the balance between the processes of addition of organic materials (plant residues, among

Treatments	N		P		K	
	Without lime	With lime	Without lime	With lime	Without lime	With lime
			$g\,kg^{-1}$			
Controls	12.5 Bb	14.7 A	0.83	0.89	4.48	5.04
Cupuassu shell	14.9 Aab	13.9 A	0.97	0.97	5.42	5.20
Cupuassu shell+ Inga	15.9 Aa	14.8 A	0.88	0.97	4.29	5.53
Mean lime	14.4	14.4	0.89 B	0.94 A	4.73 B	5.26 A

Means followed by different lowercase letters in the columns, and different upper-case letters in the lines, differ from each other, by Tukey test at 5% probability.
12 t ha^{-1} organic fertilization of residues of cupuassu shell.
22 t ha^{-1} organic fertilization of residues of cupuassu shell + 3 t ha^{-1} Inga pruning.

Table 4.
Concentrations of nitrogen (N), phosphorus (P), and potassium (K), in cupuassu tree, leaves planted in an Oxisol of Central Amazonia as a function of lime and organic fertilizer.

others) and their loss (mineralization and decomposition by the decomposing organisms present in the soil). Under the conditions this work was performed, the dolomitic liming might have stimulated the mineralization of the organic matter added to the soil in the form of plant residues, which, combined with the high temperatures and humidity moisture (rainy season), favored and accelerated the decomposing process.

Inputs of exogenous organic matter may accelerate or retard the mineralization of native soil organic carbon (SOC) through a priming effect, and thus have a potential to change SOC dynamics [25]. In general, the priming effect is induced by the exogenous organic C but its intensity is controlled by soil nutrient availability [26]. Under the conditions of this work, liming increased the mineralization of N, increasing the availability of this nutrient in the soil (**Figures 1** and **2**), which probably accelerated the decomposition of the native SOC, [26, 27]. However, more research is needed to clarify the influences of organic amendments on SOC build-up [28].

3.3 Mineral nitrogen contents (N-NH_4^+ and NO_3^-)

Figure 1 shows the results of the mineral N, in the soil, as nitrate (N-NO_3^-) and ammonium (N-NH_4^+). The predominant mineral N was the form of N-NH_4^+, especially in the treatment with the addition of Inga pruning, with presented this nutrient's highest releasing rate (0–5 cm layer) 60 days after its application. Mineral N predominant in the form of ammonium (N-NH_4^+) confirms the findings by other authors who claimed this ion to be a major N source for plants on Amazonian Oxisol [29]. The lower content of N in the form of nitrate (N-NO_3^-) might result from its greater losses through leaching and denitrification, such as documented by other authors [30, 31] and by its use by plants. On the other hand, the presence of the two forms of nitrogen in the soil can be highly positive for plant nutrition to maintain the internal ionic balance, since their uptake is in the form of N-NH_4^+ (positive charge) or in the form of N-NO_3^- (negative charge) keep the electrical equilibrium.

Figure 2 shows the liming positive effect on the organic fertilizer mineralization 60 days following their application. The three depths showed to have occurred an increase in the total mineral N (N-NH_4^+ + N-NO_3^-) contents, confirming the data shown in **Figure 1** and adding the information of it that it was leached down to the depth of 20 cm, at least. At 60 days after the application of organic fertilizers, the mineral N total (NH_4^+ + NO_3^-) contents along the soil profile were higher in the treatments containing liming and cupuassu shells either with or without Inga residues, because adding

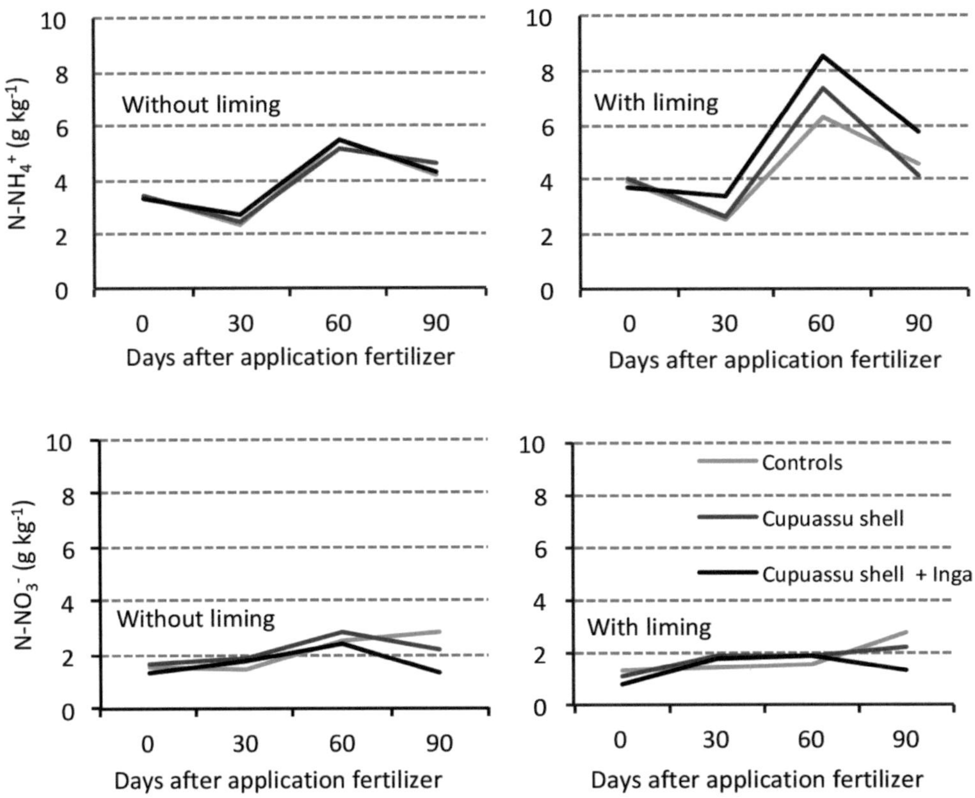

Figure 1.
N mineral content in the form of NH_4^+ and NO_3^- in an Oxisol, cultivated with cupuassu tree, as a function of lime and organic fertilization.

that organic material and raising the pH as well as the addition of Ca and Mg through dolomitic limestone increased the nitrogen contents through soil layer (**Figure 2**). It must take into account that part of this nutrient was uptake by the plants, or its contents would have shown to be higher in the soil profile due to leaching. The results obtained in this work confirm that the applications of green manure (leguminous plant prunings) are able to increase the mineralizable N in cupuassu crops in Oxisols.

3.4 Cupuassu plant nutrient contents

Liming significantly increased the plant absorption of P, K (**Table 4**). On the other hand, in the plot without liming occurred a significant increase in the absorption of N in the treatment with *Inga + cupuassu shell* compared to the control. Probably, the incorporation of organic waste has contributed to increasing the mineralization of organic N in the soil, as has been reported in other studies [29, 32]. Studies on the Oxisol and Ultisol of Central Amazonia have mentioned the absence of N response in cupuassu fruit production, due to the high mineralization rate and, consequently, high availability of mineral N in these soils, especially in soils with leguminous cover [33, 34].

The average concentrations of N and P obtained in this work are below the levels found by other authors in cupuassu tree plantations in the Oxisols of Central Amazonia [33], while the average concentration of K is found well above the values obtained in other works in Amazonia [1, 33], which may be related to the effect of organic fertilizers on the supply of K to the cupuassu plants in the present work. On Cambisols of Central Amazonia, Ayres and Alfaia [1] observed that liming promoted a small increase in K uptake by cupuassu plants.

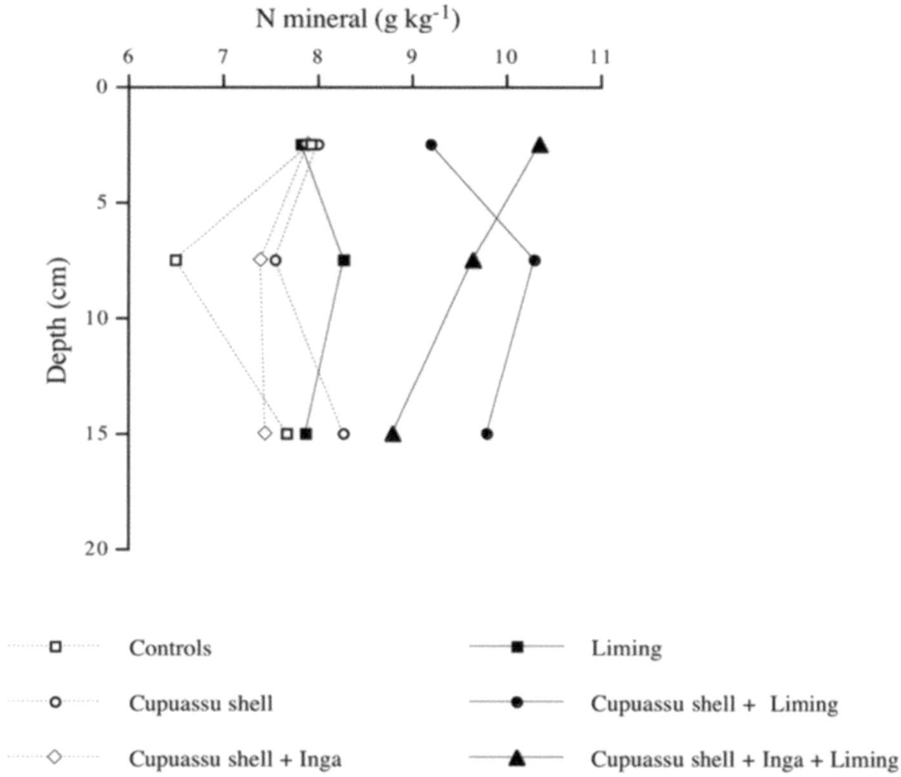

Figure 2.
N mineral total (NH_4^+ + NO_3^-) content in a Oxisol, cultivated with cupuassu tree, as a function of lime and organic fertilization.

3.5 Fruit production

The data in **Figure 3** showed that organic fertilization induced a slight, non-significant increase in fruit production. Alfaia et al. [9], in experiments using cupuassu bark residues with and without liming, also did not observe significant effects on fruit production during the first fruit harvest; however, the increase in production was most pronounced during the harvest of the following year. It is probable that in this work, the time after the application of the treatments was not enough for the decomposition of organic matter to occur.

The largest increases of nitrogen in the soil occurred at 60 days after the application of fertilizers, shown a slow decomposition of organic material (**Figure 3**), even in the treatment with the addition of Inga, which has a lower C/N ratio and could help to accelerate the decomposition process [35]. The Inga, although being a legume, has low rates of decomposition and release of nutrients, when compared to other legume species [36]. In an Oxisol in the Central Amazon, Schwendener et al. [37] observed that the mixture of cupuassu litter (of low nutritional quality) and leaves of Inga (of slow decomposition) did not contribute to the increase of mineral N in the soil in the short term, in contrast to a legume such as gliricidia (*Gliricidia sepium*), of rapid decomposition. The results of this work show the potential of organic fertilization with cupuassu shell + Inga pruning as a supplier of nutrients for the cupuassu plants. However, more studies are needed, both on the effect of the application of residues in the long term, and tests of doses and mixing with other legumes and its use in the production of biochar and composting, which would prevent the immobilization of nutrients.

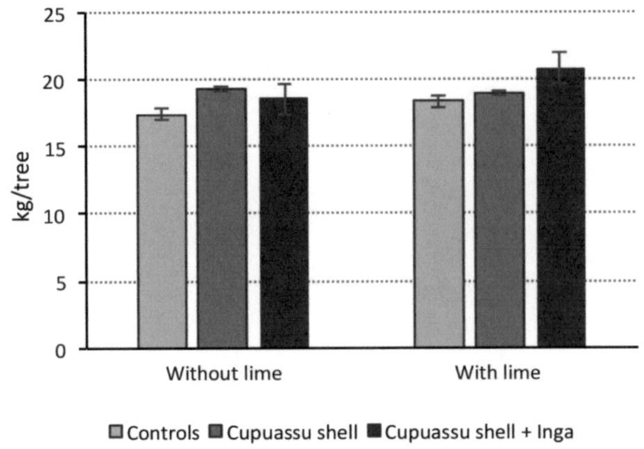

Figure 3.
Fruit production of cupuassu tree as a function of lime and organic fertilizer. Columns represent the mean of four replications and lines the standard error of the mean.

4. Conclusions

The organic fertilization with *cupuassu shell + Inga pruning* improved the chemical characteristics of the soil, mainly for the replacement of K and Ca, since applied with liming.

Without liming, there was a significant increase in the absorption of N by the cupuassu plants, showing that applications of green manure (legume pruning) can increase the N mineralization in Oxisols.

Liming also stimulated the mineralization of the organic fertilizers added to the soil, bringing about significant increases in P and K uptake by the cupuassu trees.

The assessed organic residues sources can result in a great nutrient-bearing organic fertilizer and become a low-cost alternative for recycling cupuassu processing residues.

Acknowledgements

We thank the National Technological Development Council (CNPq) and the Amazonas State Research Support Foundation (FAPEAM) for their financial support.

Conflict of interest

The author declares no conflict of interest.

Author details

Eleano Rodrigues da Silva[1], Marta Iria da Costa Ayres[2], Acácia Lima Neves[3], Katell Uguen[4], Luiz Antonio de Oliveira[2] and Sonia Sena Alfaia[2]*

1 Federal Institute of Amazonas, Manaus, AM, Brazil

2 National Institute of Amazonian Research, Manaus, AM, Brazil

3 Nacional Institute of Colonization and Agrarian Reform, Manaus, AM, Brazil

4 State University of Amazonas, Manaus, AM, Brazil

*Address all correspondence to: sonia.alfaia@inpa.gov.br

References

[1] Ayres MIC, Alfaia SS. Efeito da calagem e do potássio no cupuaçuzeiro (*Theobroma grandiflorum*) em Cambissolos da Amazônia Ocidental – características químicas no solo e na planta. In: Noda H, Souza LAG, Silva Filho DF, editors. Pesquisas Agronômicas para a Agricultura Sustentável na Amazônia Central. 1st ed. Manaus: Editora do INPA; 2013. pp. 87-100

[2] Said MM, Oliveira LA, Rivas AAF. Cultural aspects and potential use of cupuassu in the Itacoatiara county, Amazonas State. Revista de Ciências Agrárias. 2013;**56**:30-36. DOI: 10.4322/rca.2013.077

[3] Alves RM, Resende MDV, Bandeira BS, Pinheiro TM, Farias DCR. Avaliação e seleção de progênies de cupuaçuzeiro em Belém, Pará. Revista Brasileira de Fruticultura. 2010;**33**:204-212

[4] Lima HE, Santos VA, Chagas EA, Rodriguez CA, Araújo MCR. Severidade da vassoura de bruxa em genótipos de cupuaçuzeiro cultivados em sistema agroflorestal (SAF's) e produção de genótipos tolerantes a doenças. Cadernos de Agroecologia. 2013. Available from: http://revistas.aba-agroecologia.org.br/index.php/cad/article/view/14378. [Accessed: August 14 2018]

[5] Dias BO, Silva CA, Higashikawa FS, Roig A, Sánchez-Monedero MA. Use of biochar as bulking agent for the composting of poultry manure: Effect on organic matter degradation and humification. Bioresource Technology. 2010;**101**:1239-1246

[6] Lopes ELN, Fernandes AR, Teixeira RA, Sousa ES, Ruivo MLP. Soil attributes under different crop management systems in an Amazon Oxisols. Bragantia. 2015;**74**:428-435

[7] Tomazeli AN, Santos I, Morales RGF. Resíduos orgânicos para o controle das doenças do feijoeiro causadas por *Sclerotium rolfsii*. Revista Ambiência. 2011;**7**:65-74

[8] Dunaj SJ, Vallino JJ, Hines ME, Gay M, Kobyljanec C, Rooney-Varga JN. Relationships between soil organic matter, nutrients, bacterial community structure, and the performance of microbial fuel cells. Environmental Science & Technology. 2012;**46**:1914-1922. DOI: 10.1021/es2032532

[9] Alfaia SS, Silva NM, Uguen K, Neves AL, Dupin B. Pesquisa participativa para recuperação da produtividade de sistemas agroflorestais na Amazônia Ocidental: o caso do Projeto Reca, Nova Califórnia, RO. In: Porro R, editor. Alternativa Agroflorestal na Amazônia em Transformação. Brasília, DF: Embrapa Informação Tecnológica; 2009. pp. 781-804

[10] Dias JRM, Wadt PGS, Perez DV, Silva LM, Lemos CO. DRIS formulas for evaluation of nutritional status of cupuaçu trees. Revista Brasileira de Ciência do Solo. 2011;**35**:2083-2091

[11] Silva FC. Manual de análises químicas de solos, plantas e fertilizantes. 2nd ed. Brasília, DF: Embrapa Informação Tecnológica; 2009. 627 p

[12] EMBRAPA. Manual de métodos de análise de solo. Rio de Janeiro: Embrapa/CNPS; 2011. Available from: https://ainfo.cnptia.embrapa.br/digital/bitstream/item/104933/1/Manual-de-Mtodos-de-Analise-de-Solo.pdf [Accessed: August 14 2018]

[13] Moreira A, Fageria NK. Soil chemical attributes of Amazonas State, Brazil. Communications in Soil Science and Plant Analysis. 2009;**40**:1-14. DOI: 10.1080/00103620903175371

[14] Franchini JC, Gonzalez-Vila FJ, Cabrera F, Miyazawa M, Pavan MA.

Rapid transformations of plant water-soluble organic compounds in relation to cation mobilization in an acid Oxisol. Plant and Soil. 2001;**231**:55-63

[15] Castro FLPC. Produção de compostos orgânicos com resíduos de pirarucu (*Arapaima gigas* Schinz.) associados a outras fontes orgânicas no município de Maraã AM. 2017 [master dissertation]. Manaus: Instituto Nacional de Pesquisas da Amazônia; 2017

[16] Bessho T, Bell LC. Soil solid and solution phase changes and mung bean response during amelioration of aluminum toxicity with organic matter. Plant and Soil. 1992;**140**:183-196

[17] Pinho RC, Miller RP, Alfaia SS. Agroforestry and the improvement of soil fertility: A view from Amazonia. Applied and Environmental Soil Science. 2012; Article ID 616383, p. 11. DOI: 10.1155/2012/616383

[18] Sanchez PA, Bandy DE, Villachica JH, Nicholaides JJ. Amazon basin soils: Management for continuous crop production. Science. 1982;**216**:821-824

[19] Suominen L, Ruokolainen K, Tuomisto H, Llerena N, Higgins MA. Predicting soil properties from floristic composition in western Amazonian rain forests: performance of k-nearest neighbour estimation and weighted averaging calibration. Journal of Applied Ecology. 2013;**50**:1441-1449. DOI: 10.1111/1365-2664.12131

[20] Grau O, Peñuelas J, Ferry B, Freycon V, Blanc L, Desprez M, et al. Nutrient-cycling mechanisms other than the direct absorption from soil may control forest structure and dynamics in poor Amazonian soils. Scientific Reports. 2017;**7**:45017. DOI: 10.1038/srep45017

[21] Alfaia SS, Uguen K. Fertilidade e Manejo de Solos. In: FMS M, Cares JE, Zanetti R, Stumer SL, editors. O Ecossitema Solo - Componente, relações ecológicas e efeitos na produção vegetal. Lavras, MG: Editora da UFLA; 2013. pp. 75-90

[22] Wandelli EV, Ferreira F, Souza GF, Souza SGA, EKM F. Exportação de nutrientes de sistemas agroflorestais através de colheitas. O valor dos resíduos dos frutos amazônicos. In: Anais do VI Congresso Brasileiro de Sistemas Agroflorestais, 25-28 October 2002; Ilhéus, BA, Brazil. SBSAF 2002. Congress proceedings. Available from: https://www.alice.cnptia.embrapa.br/alice/bitstream/doc/672115/1/7028.pdf

[23] Alfaia SS, Uguen K, Rodrigues MRL. Manejo da Fertilidade dos Solos na Amazonia. In: FMS M, Siqueira JO, Brussaard L, editors. Biodiversidade do solo em ecossistemas brasileiros. Lavras, MG: Editora da UFLA; 2008. pp. 117-141

[24] Khorramdel S, Koocheki A, Mahallati MN, Khorasani R, Ghorbani R. Evaluation of carbon sequestration potential in corn fields with different management systems. Soil & Tillage Research. 2013;**13**(3):25-31. DOI: 10.1016/j.still.2013.04.008

[25] Fontaine S, Bardoux G, Abbadie L, Mariotti A. Carbon input to soil may decrease soil carbon content. Ecology Letters. 2004;**7**:314-320

[26] Chen R, Senbayram M, Blagodatsky S, Myachina O, Dittert K, Lin X, et al. Soil C and N availability determine the priming effect: Microbial N mining and stoichiometric decomposition theories. Global Change Biology. 2014;**20**:2356-2367. DOI: 10.1111/gcb.12475

[27] Qingyan Q, Lanfang W, Binbin L. Crop residue-derived dissolved organic matter accelerates the decomposition of native soil organic carbon in a temperate agricultural ecosystem. Acta Ecologica Sinica. 2019;**39**:69-76. DOI: 10.1016/j.chnaes.2018.05.006

[28] Yehong X, Zengming C, Weixin D, Jianling F. Responses of manure decomposition to nitrogen addition: Role of chemical composition. Science of the Total Environment. 2017;**587-588**:11-21. DOI: 10.1016/j.scitotenv.2017.02.033

[29] Alfaia SS, Jacquin F, Guiraud G. Transformation of nitrogen fertilizers in Brazilian Amazonia soils. Arid Soils Research and Rehabilitation. 1995;**9**: 335-340

[30] Masclaux-Daubresse C, Daniel-Vedele F, Dechorgnat J, Chardon F, Gaufichon L, Suzuki A. Nitrogen uptake, assimilation and remobilization in plants: challenges for sustainable and productive agriculture. Annals of Botany. 2010;**105**:1141-1157. DOI: 10.1093/aob/mcq028

[31] Liu C-W, Sung Y, Chen B-C, Lai H-Y. Effects of nitrogen fertilizers on the growth and nitrate content of lettuce (*Lactuca sativa* L.). International Journal of Environmental Research and Public Health. 2014;**11**:4427-4440. DOI: 10.3390/ijerph110404427

[32] Alfaia SS, Guiraud G, Jacquin F, Muraoka T, Ribeiro G. Efficiency of 15N-labelled fertilizers for rice and ryegrass cultivated in an Ultisol of Brazilian Amazonia. Biology and Fertility of Soils. 2000;**31**:329-333

[33] Schroth G, Elias MEA, Macêdo JLV, D'angelo AS, Liberei R. Growth, yields and mineral nutrition of Cupuaçu (*Theobroma grandiflorum*) in two multi-strata agroforestry systems on a ferralitic Amazonian upland soil at four fertilization levels. Journal of Applied Botany. 2001;**75**:67-74

[34] Alfaia SS, Ayres MIC. Efeito de doses de nitrogênio, fósforo e potássio em duas cultivares de cupuaçu, com e sem semente, na região da Amazônia Central. Revista Brasileira de Fruticultura. 2004;**26**:320-325. DOI: 10.1590/S0100-29452004000200033

[35] Parton W, Silver WL, Burke IC, Grassens L, Harmon ME, Currie WS, et al. Global-scale similarities in nitrogen release patterns during long-term decomposition. Science. 2007;**315**:361-364. DOI: 10.1126/science.1134853

[36] Gallardo-Ordinola JLE. Produção de liteira em sistemas agroflorestais implantados sobre pastagens degradadas e sua influência sobre características bioquímicas do solo. [master dissertation]. Manaus: Instituto Nacional de Pesquisas da Amazônia/Universidade Federal do Amazonas; 1999

[37] Schwendener CM, Lehmann J, Rondon M, Wandelli E, Fernandes E. Soil mineral N dynamics beneath mixtures of leaves from legume and fruit trees in Central Amazonian multi-strata agroforests. Acta Amazonica. 2007;**37**:313-320. DOI: 10.1590/S0044-59672007000300001

Chapter 9

The Insects as a Workforce for Organic Fertilizers Production – Insect Frass

Regina Menino and Daniel Murta

Abstract

Following the evolution of composting technology, the process of digestion of a biological substrate by insects (entomocomposting) represents the last stage; however, from its initial context of producing an organic fertilizer, the role of entomocomposting has been imposing itself (due to increasing demographic pressure) mainly in the safe disposal of organic waste (in rampant growth) and in the breeding of insects for food and feed, for the sake of food security. Both these last goals converge in the first, as the safest disposal of the compost is its use as organic fertilizer; but the organic substrates are of a diversified nature, as are the species of insects which have already proved themselves in entomocomposting; hence, for each of the purposes in view, the choice is vast and, in the same way, the entomocompost composition is wide-ranging. Furthermore, various types of organic substrates, in addition to a microbial flora with symbiotic effects, may sometimes be able to transmit to the frass a harmful load of heavy metals and/or, depending on the composting insect agents, the presence of microorganisms harmful to crops and to humans and animals; in these situations, the former should be encouraged, and the latter counteracted through appropriate composting technology. Directives and legislation in this area, if properly considered, constitute a fundamental basis for ensuring the appropriate use of this particular kind of organic fertilizer. Apart from the production of insects for food and feed, where the choice of which insect is determined at the outset, the preference for the insect to be used in entomocomposting should be considered according to its proficiency in biological digestion of the organic substrates available for this purpose and the fertilizing quality of the frass produced. Although a multitude of species have been evaluated, to date, for the digestion of organic substrates, most have been used in assessing their specific potential for certain functionalities of frass related to crop nutrition and health, but there are few which, either by prolificacy, proficiency or rapidity in digesting substrates, exhibit capacity to compete in rural environment; nevertheless, new species could be evaluated in the framework of the research of competitors for entomocomposting of all or each substrate type and for each of the main anticipated objectives, meanwhile, genetic improvement to obtain new strains specialized for different organic substrates has already started to take its first steps. In addition to the binomial "insect x substrate" the composting technology constitutes the third fundamental factor for the efficiency of the process. Insects use as a composting agent has been suggested several decades ago, but it was only in the last decade that this process grown from the garden to the factory. Within rural areas, entomocomposting could play a key role within a circular economy, where recycling and reusing potentially polluting wastes safely returns to the land the enduring fertility that enables the sustained production that generated them, requiring no particularly upscale installations, equipment or technical training; it can, therefore, be adapted to any size of agricultural holding, from smallholdings to large industrial holdings, on the other hand, and in order to obtain a controlled production and high quality entomocompost, it is needed to implement industrial technologies and the composting unit can achieve a very high production per square meter, comparing with traditional composting methods. However, whether from the perspective of agriculture, livestock or forestry, the production of waste for entomocomposting always falls far short of the necessary scale, and therefore always requiring the use of biodigested organic waste from agricultural industries, provided that the necessary precautions are taken; in any case, it always constitutes added value, due to the products it generates, in addition to the inestimable value of the productive disposal of potentially polluting products. Despite all the advantages mentioned above, the controversy over the organic vs. mineral fertilizer option persists, often fuelled by myths on both sides, but the successes already achieved with insect entomocomposts, such as the black soldier fly (*Hermetia illucens* L.) or the mealworm (*Tenebrio molitor* L.), in field trials, which are gradually adding up, anticipate an important role for insects in safeguarding global food and environmental security.

Keywords: soil fertilizers, biodigestion, entomocompost, insect frass, sustainable agriculture

1. Introduction

The first alternative to the traditional aerobic composting of organic waste was vermicomposting; however, according to Čičková *et al.* [1], Lindner already in the second decade of the last century (1919) had proposed the use of insects (in accordance to his experiments with the house fly (*Musca domestica* L), to recover nutrients (especially fat) from organic waste; this could have been the threshold for entomocomposting research, nevertheless, it is only a few decades later that some experimental work appears in the context of the disposal of potentially polluting organic waste.

Nowadays, suitable technologies have already been developed for a diverse range of target functionalities of entomocomposting, namely:

- The treatment of polluting organic waste for safe disposal, from the point of view of environmental safety;
- The production of insects (in the larval, pupal or imaginal stage), from the perspective of food and feed security;
- and the production of an organic fertilizer, from the perspective of rationalizing sustained agricultural production and reducing to minimum the environmental impact of agricultural activity, without prejudice of the optimum levels of the potential production in each "soil-climate-culture" interaction process.

Whatever the main objective is, the entomocompost will always provide invaluable added value (as produced or, possibly, after suitable treatment) for safeguarding food security and the environment, due to its potential in the immediate and deferred fertilization of agricultural soils, so as in the recovery of exhausted soils and even in the inclusion of new unproductive land, as, for example, in the case of abandoned mining sites.

But when we talk about entomocompost we are referring to a product with a very diversified physical, chemical and microbiological composition, depending on the nature of the substrate to be digested and the species used to produce it; thus, the option to obtain an organic fertilizer by entomocomposting for pre-defined main purposes, requires taking the following into consideration, as regards the elements of the above mentioned "substrate-insect" interaction: On the one hand, knowledge of the nature of the substrates to be biodigested and the more adequate species to do the job; on the other hand, the eventual treatment of the substrate that may enhance more efficient bio-digestion and/or specific qualities of the frass, and the choice of the best insect species (or genotype within the elected species) in order to obtain most efficiently the entomocompost that is more suitable for fertilizing the soil and the crop for which it is intended.

This would be ideal, and to some extent feasible, at rural level, however, the production of entomocompost for crop fertilization is still in short supply, even counting on the compost obtained in a different context.

Organic bio-digestion by insects for the production of entomocompost also requires the choice of technology to be adopted, both for industrial production and for medium-scale production in rural areas (either at cooperative scale or at farm level), and in the case of agricultural and livestock holdings, it evokes the choice of a circular economy system.

This chapter deals with the knowledge that informs the decision to be taken for all the aforementioned options, by the some other, concluding with a compilation of the experimental results that we consider most relevant with regard to the relative fertilizing potential (immediate and deferred) for specific situations of the "soil–plant-fertilizer" triad interaction. Some notes on insects regarding food and feed are anticipated, as they are inextricably linked to the production of the entomocompost.

2. Insects as food and feed

The present COVID-19 pandemic has shown how Europe is hostage of the international feed market, and as far as nutrition is concerned, protein is a huge problem to be solved. However, society continues to waste food products, contributing to a very inefficient agriculture vale chain in which more than 25% of food products can be lost.

Based on a one hundred percent circular economy-based approach, vegetable by-products can be converted into high valuable nutrient sources for both animals and plants. Insects can be the key for the transformation of this otherwise lost nutrients into new nutritional solutions not only for both humans and animals, but also for plants.

In a very short period, insects can convert a very large range low value by-products into high value insect protein and oil for animal and human nutrition and insect frass, and organic fertilizer for plants. With this process, now completely industrialized and at a full-scale level, it is possible to reduce the Europe dependency from the international feed and food markets, contributing to a local and more sustainable food production.

As mentioned, nowadays, feed producers face several significant global challenges to find suitable resources to produce compound feed for livestock, aquaculture, and pets. On the one hand, the growing demand for animal products, and thus for animal feed, associated with the need to find resources with reduced environmental impact, has led to the development of novel feed ingredients, and moves to decrease dependency on common resources, such as soybean meal, maize and fishmeal. The current use of these resources is assessed as being unsustainable therefore driving the need for alternative ingredients to maintain the balance between food, feed and biofuel industries. Land degradation, water deprivation and drastic climate change are additional challenges impacting on livestock production, aquaculture and the pet food industry.

On the other hand, recent events have illustrated the need to reduce our dependency on imported resources, specifically from other continents, strengthened by consumer opinion exerting pressure to provide more 'natural' food production for humans, livestock and pets. Accordingly, the development of novel sustainable raw materials plus improved efficiency of resource use play, and will continue to play, a vital role in ensuring the sustainability of feed manufacture.

Significant relevance is now placed on the development of new feed resources based on environmentally friendly approaches, circular economy solutions, and the use of natural resources. However, it is not likely in the near future to be feasible to completely replace existing feed ingredients with novel ones, leading to a focus in the sector on improving efficient use of existing ingredients, thus decreasing demand.

Some novel food and feed ingredients have the significant advantage of making use of available agri-food co-products and transforming them locally into new nutrient sources. Insects are one such ingredient that has the capability to convert low value vegetal co-products into a high value nutritional solution, while also aligning with the environmental drivers that are prompting the food and feed revolution.

It has been estimated that food waste accounts for 23% of arable land and 24% of freshwater resources used for crop production [2]. Thus, it is relevant to evaluate the use of insects in feed from a circular economy point of view. Insect rearing can potentially be used to upgrade low-value organic food waste streams increasing the efficiency of natural resource use and animal production.

Several livestock production companies in the world operate vertically, producing feed for the animals, raising and processing them before market. Co-products include manure and other animal and vegetable co-products as well as former foodstuffs. Insects could be an invaluable tool for such organizations, as they can provide a perfect link between nutrient loss in vegetable co-products, and the protein supplement needed for animal feed.

Therefore, insects have a perfect spot in certain value chains, where, more than creating value, they contribute to natural resource use efficiency through nutrient bioconversion. This might be the greatest contribution of insects to the food value chain, as they have the capability to be integrated perfectly in present day market chains, whilst also converting wastes and less desirable co-products into high value nutrient resources. When applied with the right infrastructure, such systems could contribute to animal production efficiency, environmental sustainability, and supply chain profitability. Furthermore, insects, as for other novel food and feed ingredients, offer the potential to decrease dependency upon foreign products imports, creating new local products, and thus helping to shorten supply chains.

Thus, there is growing interest regarding the use of insects as an alternative ingredient source for both food and feed production. The use of alternative ingredients in animal diets can be optimized in terms of their nutritional characterization, their safety and technological quality, in order to achieve better performance as well as facing the challenges of increased feed demand in volume as well as quality and sustainability factors.

Insects can supplement traditional feed sources such as soy, maize, grain and fishmeal, with several different species of insects considered for use as a partial or total substitute of traditional feed sources [3–6]. Many trials have been conducted with different animal species, both terrestrial and marine, with the challenges associated with the use of insects in these animals changing, dependent not only on the animal species being fed, but also on the insect species being used, and the rearing substrate on which it was grown. However, it has also been demonstrated that different organic substrates can be used to rear insects, such as Black Soldier Fly (BSF), without significantly affecting its amino acid composition, the profile of which has been shown to be similar to that of fish meal and soybean meal [3, 7]. By contrast, when considering fat and ash composition, both can differ substantially according to the rearing substrate [8]. Thus, insect nutritional and technological properties are linked to the species, rearing system adopted and especially to the substrate used [8].

On the other hand, the so produced novel plant nutritional source, entomo-compost, can contribute to a wide range of soil solutions, from drought resistance and plant nutrition to even pest control and sprouting promotion.

However, this novel sector still faces several challenges, from legal to consumer acceptance and to industrialization and growth. Although the legal framework is changing and adapting to this new reality, consumers still have to prepare for it, and insect producers have a lot to learn from other livestock and industrial sectors. Besides that, the use of insects as a tool to other applications is still in its infancy, as insects can be used from bioremediation in garbage disposal systems, to the production of new plastic solutions.

3. On the substrate to be composted

If initially the use of uncontrolled composting of organic agricultural wastes may have been motivated by obtaining a fertilizer for crops, as in the case of the use for this purpose of animal bedding and manure slurry, the fact is that subsequently, with the exponential growing production of industrial and urban organic waste, the emphasis has shifted to the disposal of polluting waste.

Within this last context, controlled composting methods have been developed [9], including in its motivation for the production of arthropods and worms for food and feed, and not least for the production of organic fertilizers; in fact, food security is no less relevant, and to this end the resilience of agricultural soil fertility, the restoration of depleted soils and even the acariation for agricultural production of infertile soils hitherto ignored for this purpose, without compromising environmental security, is urgently needed.

But fertility is an ambiguous concept when applied in relation to the productive capacity of agricultural soils. Actually, boosting the full productive capacity of a given plant species (or even of a given genotype of the same species), depends on rigorously reconciling its physical and chemical requirements with the soil and climate conditions in which it is located, and these can be very diverse – in poor soil conditions, a primitive variety has more yield potential than a variety improved for yield capacity [10, 11].

However, that above-mentioned objective, of full productive capacity, will not always be the most appropriate if it is not based on economic, environmental and ethical considerations, because maximizing production does not always lead to greater financial return, it frequently translates into an environmental burden and it often forgets the responsibility of the agricultural sector in the context of global food security.

With regard to soil fertilization, which underlies the subject of this Communication, the above considerations are also valid: Soil fertilizers should be required to provide an advantageous cost–benefit balance, to cooperate in protecting the environment, and to increase the resilience of soil fertility as a basis for long-term food security. According to these requirements there seems to be difficult to find a perfect type of fertilizer.

As with the majority of organic fertilizers, the main virtue of entomocompost lies in their action in correcting the physical, chemical and microbiological properties of soils (a fundamental factor in their deferred fertility) and in supplying the mineral elements necessary for each crop, in each specific situation (an important factor for their immediate fertility) [12]. Furthermore, entomocomposting can be a relevant factor in recycling exhaustible plant nutrient resources; an example of this is mentioned by Zhang *et al.* [13], when they observe the accumulation of phosphorus in the frass of grasshoppers as a function of the stoichiometric homeostasis of the N:P ratio in their bodies.

Any entomocompost is, however, more than the frass obtained from a given substrate; apart from the frass, it contains the remainder of the substrate and residues from the metamorphosis of pupae into adults (if not the pupae themselves). This is not entirely true in the case of biological digestion by the BSF larvae, which evidences the unique behavior of abandoning the compost at the pre-pupa stage, a phenomenon with obvious advantages in pupal harvesting and which is referred to in the bibliography as self-harvesting. For this reason, for the entomocompost obtained from BSF was proposed [14] the acronym CASH (Compost After Self Harvesting).

But the constraints on the more generalized use of entomocomposts are not limited to those mentioned above. They range from lack of definition of the exact formulation of their composition, to lack of knowledge of the mineralization rate (for formulations of its fertilizing elements that can be directly assimilated by the plants) in the soil and climate situations in question, to logistical and scale limitations to supply. In reference to this latter setback, Timsina [15] states that "considering the current organic sources of nutrients in developing countries, organic nutrients alone are not enough to increase crop yields to meet global food demand".

Through bio-digestion by insects, the formulation of the substrate is largely altered in its physical, chemical and microbiological composition, with decisive consequences on the fertilizing potential of the entomocompost; it, therefore, plays a relevant role in the quality of the compost. As referred by Poveda *et al.* [16], by modifying the insect diet, not only do you get different nutritional content in the frass, but also significant changes in the actual microbiota, both aspects relevant to its ability to be used as organic fertilizer.

As an extreme situation, regarding the nature of the substrate, Koh *et al.* [17] reports that polystyrene, when digested by the coleopteran *Zophobas morio*, produced a starch-rich frass, which promoted the growth of *Hylocereus undatus* plants from both the aerial and root parts.

An entomocompost of a substrate of agroindustry origin, or of remnants and residues from agricultural production, is not seen as a threat of chemical or microbiological contamination of agricultural soil. On the other hand, a compost obtained by insect bio-digestion in industrial urban waste plants, requires analysis and possible remediation if chemical and microbiological substances harmful to soil fertility are found to be present, as for instance in the case of houseflies, with high levels of lead and arsenic in the frass [18].

Entomocompost derived from manure and slurry does not normally pose the danger of soil contamination of any kind, however, in the context of insect production for feed or for food, is not at all suitable, and is subject to severe restrictions. To overcome these constraints, which are mainly dictated by the nature of the substrates, some progress has already been made.

Thus, although the use of unsafe wastes as substrate for entomocomposting can be done with efficiency, as this technic is mainly applied for food and feed purposes, it is quite uncommon to see full-scale insect rearing units using such substrates. In fact, the majority of entomocomposting units, or insect farms, are using vegetable coproducts as substrates for insect production, being its main purpose the production of insects as a protein source and the insect frass a co-product of this process. Nevertheless, and as previously demonstrated, entomocompost is a high valuable product with very good effects in soil fertility and plant health.

This way, entomocomposting should not only be considered as a process to produce food and feed, but also to produce the entomocompost as a main objective, opening the use of other, unsafe, wastes that cannot be used when the produced insects are intended for the food chain. However, such approach will require studies to evaluate food safety concerns in a one health approach, evaluating from environmental impacts and benefits, to possible impacts in the soil and plants. The so produced insects cannot be intended for the food cahin but might well find an economic value as a raw material for biorefineries such as fuel and plastic, or to be also used as fertilizers.

The use of entomocomposting technics to convert unsafe substrates such as urban organic wastes, manure or sewer sludge, is being tried in several R&D projects. Such is the case of NETA project (POCI-01-0247-FEDER-046959), a project in which a new manure and sewer water is being treated with a novel process and the sludge is being tested as a substrate for entomocomposting. This produced insect frass is being tested in vegetable and olive oil production, while the larvae are being evaluated in terms of safety, evaluating both chemical and microbiological contaminants, and being used for industrial purposes.

In order to be possible to produce insects for food and feed purposes with organic wastes as substrate, what would unlock the entomocomposting potential as a bioremediation tool, one should first show if such approach is safe. However, before proving its safety, one of the main challenges in entomocomposting organic wastes is that if we used the same insect species as for the production of food and feed, it will not be possible to differentiate insect products produced with safe or unsafe substrates. Thus, one possibility would be to develop the entomocomposting process of unsafe substrates, such as organic wastes, manure and sewer sludge, with insect species not being used for food and feed purposes. That would allow to differentiate the obtained insect product with DNA testing and would unlock a very beneficial tool for the treatment of high environmental impact organic wastes, transforming them into novel products and returning lost value to the value chain, while contributing to both economic growth and sustainability in a 100% circular economy approach.

Thus, it should be highlighted that using insects for nutrient production is not a goal in itself but can be an instrument to achieve goals in biowaste reduction and conversion, improving sustainability and optimizing the food value chain. Insects should be evaluated as a tool to increase the efficiency of resources use and to increase income, and thus, one must evaluate them beyond their nutrient value as a feed ingredient.

For example, BSF are a rich source of lipids which can be industrially extracted to obtain a pure oil with several different potential uses, from feed and food, to biodiesel and cosmetics. It has been shown that BSF fat could be a useful alternative for other commonly used fats, with specific technological properties in common with palm and coconut oils, which are increasingly associated with negative environmental impacts. In particular, the melting and crystallization behavior of BSF larval fat seems to allow replacement for traditional fats [19].

In addition, the insect exoskeleton can be processed to obtain chitin and chitosan, and its industrial scale production could offer a potential source of prebiotic oligosaccharides for pet, animal, and human nutrition [20]. Applications for chitin and chitosan go beyond nutrition, as chitosan is characterized by non-toxicity, biodegradability, film-forming capacity, antimicrobial and antioxidant properties and good barrier properties of packaging films against oxygen [21–23]. Thus, the potential for the use of insect derived chitosan to produce biodegradable plastics is being evaluated for a variety of applications, ranging from agriculture to food packaging.

Chitin-derived products have also been shown to be toxic to plant pests and pathogens, inducing plant defenses and stimulating the growth and activity of beneficial microbes. Chitin-based treatments augment and amplify the action of beneficial chitinolytic microbes [24]. Such properties prompted the development of novel crop fertilizer and crop protection products, which can be used in conjunction with one of the main insect products, the insect frass. In natural conditions, it is well known that frass deposition to soil has a great impact on soil fertility due to its high nutrient and labile carbon content and, therefore, several companies have already started to sell frass as a fertilizer [25].

4. Insects as agents to produce organic fertilizers

With all the economic and environmental advantages in the search for agricultural production that is compatible, in a sustainable way, with global demand, entomocomposts have been affirming themselves as an important alternative for reducing (if not replacing, in some cases) synthetic mineral fertilizers. To this end, several insect species have been evaluated for their proficiency in composting organic substrates.

In a careful literature review, Poveda [26] presents two thorough lists of studies on the use of insect frass as fertilizers, indicating, for each case, the plants, the benefits and the mechanisms by which these benefits were expressed.

In addition to providing the necessary mineral nutrient elements for plants, the benefits provided by adding insect frass to the soil are diverse in nature, such as: increased germination, sprouting, growth and nutritional content of plants; increased tolerance to abiotic stress; activation of the plant's defense mechanisms against pathogens and pests; increased nitrogen in plant tissue; reduced oviposition of pest insects and; increased microbial activity in soils.

However, these advantages are not all concentrated in a single species so, although frass from various species may have a relevant role as a complementary fertilizer in specific situations, few species have shown the potential to produce an entomocompost with the potential to be an alternative to avoid completely mineral fertilization in all situations. One of this specie contradicts the thesis of Lardé [27], supported by Smetana *et al.* [28], that one species cannot be suitable for the huge diversity of organic substrates - BSF has proven to be quite "cosmopolitan", living comfortably in any type of organic substrate experienced so far [29–33]. Note, however, that even in cases - in experimental situations - where entomocompost shows the potential to provide a reduction in synthetic mineral fertilizer, the fertilizing effect of entomocompost can still be enhanced if associated with appropriate soil handling technology. In this context, Dulaurent *et al.* [34], in a pot trial with frass, reports a significant increase in nutrient content in the plant by the addition of earthworms (*Lumbricus terrestris*), by promoting an acceleration of the recycling of fertilizing elements from the frass.

But if the possibility of biotic associations or of physical or chemical corrections of entomocompost for preferential purposes (within the versatility of its benefits in crop nutrition and soil fertility resilience) is a proven reality in the experimental and commercial field, in the field of genetic enhancement of insect species for entomocomposting only the first steps have been taken.

Advances in this field are predictable and particularly desirable, notably when it comes to insects with the ability to adapt to a wide diversity of substrates. This is the case of BSF, which is able to efficiently biodigest manure from various livestock species (from polygastric, monogastric or fish species) as well as residues and remnants from crops or from agroindustry.

The methodologies for this could be very diverse, but the simple continuous selection of pupae fed on the digestion of specific substrates can lead (as it has happened with the generality of animal and plant species already submitted to human-induced selection) to the differentiation of specialist genotypes more competent than generalists, probably because it should be anchored in an evolution towards more targeted physiological mechanisms that are necessarily less energy-demanding; this option, which would certainly not meet with the disagreement of the detractors of transgenics, would only require, as an additional effort, the separation of breeding facilities for flies, even though it may be slower in results, but "constant dripping wears the rock away".

5. Entomocomposting technologies

Insect production has grown a lot in the last decade. This new sector emerged with the support of FAO-UN who first referred to this field in the beginning of the decade and started an insect rush in several countries, with the development of new business. However, by then, both the business and the process where not mature yet and it took several years of development to see the first full-scale insect rearing unit being built by a handful of companies. However, the legal framework had not grown at the same rhythm, what promoted a lower growth. At this point, different companies have developed their entomocomposting technics in parallel, and even using the same insect species and substrates the processes can follow completely different approaches.

Besides that, although several approached have been made to create technologies to produce insects, and entomocompost, at a small scale, and although it can be applied at the farm level, it is only economically relevant at a large and controlled scale, ensuring both food safety and traceability.

Large scale insect production is an industrial sector in which several tons of vegetable by-products are converted by insects every day. Contrary to most composting technologies, insect production generally does not use piles of by-products. It processes them into controlled mixes of raw-materials ready to be digested. This raw material processing allows a steady rhythm of conversion and production. In most cases insects are thus reared inside plastic boxes of different sizes in large controlled environment warehouses. The time needed for composting and the number of insects to be used to convert each ton of by-products change from insect specie to insect specie and between companies. The main insect species to be used for food and feed are *Tenebrio molitor* and *Hermetia illucens* (BSF), however the last one is more prone to be used as an entomocomposting tool, as it have a large range of vegetable by-product conversion capabilities. BSF can convert decaying by-products in as few as 7 days, depending on the technology used, and some of the already existing BSF production units can convert as much as 100 tons of vegetable by-product every day into 20 tons of insect frass, while also producing 17 tons of insect larvae. However, this numbers and process greatly change between companies, which all apply different technologies, even when producing the same insect species.

Therefore, insect production has not only to achieve economically viable production at scale, investing in new full-scale insect production units, but it must also be standardized, to obtain a steady production and uniform product. Standardization is key not only in relation to a single production unit, but also between different producers. Insects as a food and feed resource, and also as a plant nutrition source, would greatly benefit from standard quality and nutritional values when considering the same insect species and product. This would increase farmers trust in this novel fertilizer. However, different insect producers may use different insect species and rearing substrates, as well as different production and processing techniques. This results in different products, with different nutritional values and properties, entering the market.

Nevertheless, as the insect rearing industry is only in its infancy, we believe that in the future the production and processing of insects and frass will tend to be more similar between operators, as different production processes and technologies attain relevance in the sector. One opportunity to increase standardization and quality of insect products might be technology transfer between companies, enabling rapid growth of this novel sector and allowing investors and new operators to enter the market without the need to invest in the development of processing technologies. Technology transfer from other companies and research institutes that have spent recent years in R&D will have processes providing the most suitable solutions, avoiding the need for new producers to start from scratch, costing time and money as well as decreasing the chances of success for new businesses.

6. The role of entomocomposting in the context of a circular economy in rural areas

For Zink & Geyer [35] "the proponents of the circular economy have tended to look at the world purely as an engineering system and have neglected the economic part of the circular economy"; to this assertion, the facts have been demonstrating, convincingly anchored in science, that the linear economy alternative, in turn, blatantly belittles the environmental part.

The circular economy is "a new economic model operating in closed circuits, catalyzed by innovation throughout the value chain" [36], and, within the agrarian economy, whether in plant or animal production, entomocomposting is an innovative alternative, more efficient than traditional composting, to reduce the import of feed and fertilizers and energy losses, with added advantages in terms of safeguarding the environment.

This is how the entomocomposting of crop remains and residues, so as livestock production wastes, is a multifaceted pivotal factor of the greatest relevance to different circularities within farms, as shown in **Figure 1**.

The circularities represented in the diagram are multiple and interlinked and are not necessarily closed. In fact, there will always be a need for outsourcing, both for supplementation of feed and fertilizer, in quantities compatible with optimizing the efficiency of the entomocompost and the feed value derived from pupae.

The protagonist in this diagram is BSF, for the peculiarities that distinguish it, in a positive way, from the other composting agents, namely:

- High prolificacy;
- High proficiency and speed in biodigestion;
- Widespread range of organic substrates that have the potential to digest;
- Efficiency in the elimination of potentially harmful microorganisms;

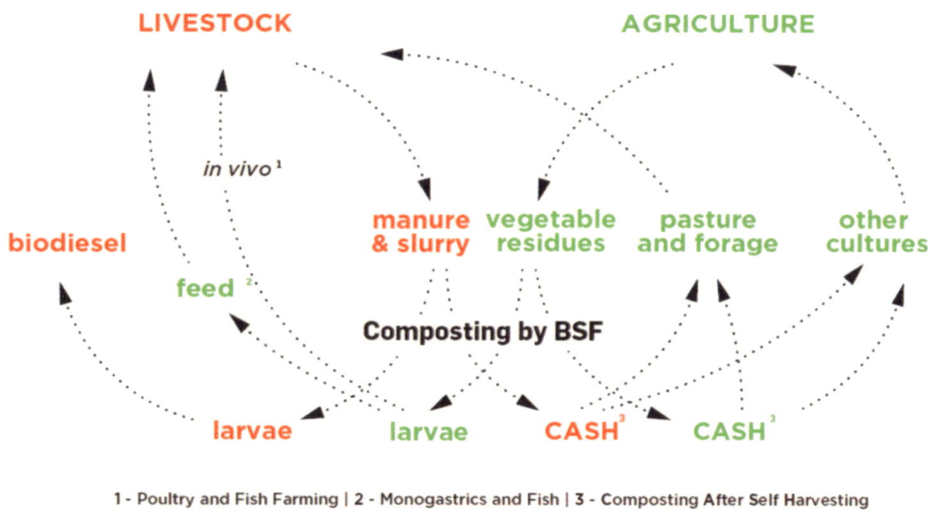

Figure 1.
Multiple circularities driven by BSF on farms.

- The ability to drastically eliminate housefly multiplication during digestion processes;
- Capacity to provide the presence of plant growth factors in the CASH.

But beyond these, two singularities, particularly relevant in agricultural and livestock holdings, distinguish them from other insects:

- Self-harvesting and;
- The fact that the adult does not have a developed mouth apparatus, feeding on the copious reserves accumulated in the larval stage, and thus does not bother (or even transmit diseases to) humans and animals.

Finally, and following the results reported by Yildirim-Aksoy *et al*. [37, 38] in trials with channel catfish (*Ictalurus punctatus*) and hybrid tilapia (*Oreocromis niloticus* x *O. Mozambique*), frass obtained by biodigestion by BSF of suitable substrates, more than an organic fertilizer, can be a feed in aquaculture.

Notwithstanding the fact that entomocomposting by BSF has already proven to be highly efficient in recycling and reuse capacity in the plant and animal production circuit, the deficit generated in the export of plant and animal products to the market means that feed and fertilizers have to be imported. To this end, in rural areas, initiatives (possibly of a cooperative nature) for the production, on an industrial scale, of entomocomposts and larvae (or pupae) for soil fertilization and feed, allow the circularity of the production system to be extrapolated from the individual sphere to the community sphere; With this type of initiative, the agroindustry will play a relevant role, with additional advantages in terms of capacity and fluidity of the system, broadening the scope of circularity at regional level.

7. Data in the context of the trinomial "soil x plant x fertilizer"

Most of the research on the role of insects in soil fertility has focused on specific aspects of the benefits of their frass, not necessarily obtained through the technological composting of organic waste but through their metabolism as part of the soil entomological fauna; with this last aim, numerous studies have been published, as it can be seen in the comprehensive listing by Poveda [26].

These *in situ* studies focus on the possible symbiotic effect of frass within the particular soil entomofauna, in different ecosystems, to assess their role in the ecological balance of the same, particularly with regard to nutrition and phytosanitary aspects of the crop species.

When it comes to the use of a compost derived from off-site insect digestion of organic waste, a more objective assessment of its fertilizing potential, although guided by scientifically well-founded theoretical considerations, should be further informed by evaluation in preliminary production trials in a conditioned environment, in accordance with the pre-defined end goal.

The fact that seldom these trials showed a decrease in production - as was reported by Alattar *et al.* [39], when comparing processed food waste via Microaerobic Fermentation and BSF larvae biodigestion as soil fertilizers in maize, or Gärttling *et al.* [40], with a BSF frass in a low nutrient potting soil - may not mean that, in the overwhelming majority of experimental trials, the results with entomocompost compete with mineral fertilizers; it should be noted that on the one hand, the implementation of field trials is often preceded by a prior study of the feasibility of the hypothesis and, on the other hand, that regrettably a large proportion of the trials that do not confirm the hypothesis are not reported.

Some pot tests have shown the potential of entomocompost, obtained from substrates of various kinds, to reduce mineral fertilization in several crops, as for instance: With mealworm (*Tenebrio molitor* L.) frass, in barley [25] and ryegrass [41], and with BSF frass, in basil and Sudan grass [42], chinese cabbage [43], yard-long bean [44], lettuce [45, 46], ryegrass [45, 47], maize [48] and swiss chard [49].

Also testing the potential fertilizer value of BSF frass from several origins and for different plant species, in pot experiment, comparing either with other organic composts or with commercial substrates, the results found by Newton *et al.* [42], Rosmiati *et al.* [50], Setti *et al.* [51] and Kawasaki *et al.* [52], were encouraging, resulting in yield increases when using frass in certain amounts.

Although these results are encouraging, the conclusive proof, which is specific to the conditions that characterize the experimental situations, can only be given in the light of the results of the field test. Trials of this nature are still scarce and will never allow abusive extrapolations, but they constitute the most valuable information on the fertilizing potential of entomocompounds in relation to mineral fertilization.

Notwithstanding the fact that Temple *et al.* [53] have not foud positive results in a field trial with beans, using a BSF EntomoCompost (BSFEC) from food waste, most of the field trials where entomocompost is used as a complement of the mineral fertilization, for contrasting with exclusive mineral fertilization, have shown an increased production for the mixt alternative. This was the case reported, among others: by Anyega *et al.* [54], in a 'acric ferralsol' trial with tomato, kale and fresh beans, with BSFEC from Brewer's spent grains; by Quilliam *et al.* [55] in a 'Ustic duraquert' trial with chili pepper and shallots, with an identical entomocompost, so as in the same soil, with maize; and by Temple *et al.* (*op. cit.*) in a 'Humic Gleysol' trial with bok choi, lettuce and potato, with BSFEC from Brewer's spent grains.

These results support the thesis that in the experimental situations tested to date, the percentage of CASH capable of competing with mineral fertilization alone, in what concerns the immediate fertilization for crops, is between 10% and 40% in volume. More optimistic results were seen, for example, in a demonstration field [56] with potato (*Solanum tuberosum*), comparing traditional mineral fertilization without and with CASH (from the digestion of agroindustrial waste of potato and onion), where a 9% increase in yield was recorded and, in addition, the tuber specific weight, and the percentage of dry matter were also higher when combining both fertilization approaches.

The arguments mentioned so far, based on experimental results endorsed in the literature, justify promising perspectives regarding the role of insects in the production of organic fertilizers capable of allowing a reduction of mineral fertilizers as far as the immediate fertilization of crops is concerned; nevertheless, more important than the immediate fertilization of crops is the deferred fertility of soils, both in the resilience or increase of their fertility and in the acariation of soils rendered unproductive by anthropogenic or climatic effects.

In any case, the medium- and long-term promotion and resilience of the fertility of the soils, which should be fostered by insect frass, would be translated, as for the generality of organic fertilizers, by the improvement of the structure of the soils and its capacity to retain water and crop nutrients and as well as by symbiotic interaction with the soil microbial flora and with the plant. Many knowledge within this perspective is still needed, but also a lot have been accumulated, allowing for hopeful evidences, such as: better use efficiency of P and K [57]; improved soil fertility and defense against pathogens [58]; suppression against *Pythium ultimum* [59]; influence on soil N availability [60]; stimulation of soil microbial activity and diversity [25]; not impairing hygienic properties of soils [47]; improvement of microbial activity [41]; increased dehydrogenase activity [61]; or increased enzyme activity (dehydrogenase and β-glucosidase) [46].

These data augur well, but medium and long-term field trials are indispensable for continued soil fertility management, since organic matter resilience is not its greatest virtue, particularly in tropical and sub-tropical climates.

Despite being still at the beginning of its career as a biodigester, for the production of organic fertilizers, and beyond the benefits of the utmost relevance in the perspective of safeguarding the environmental balance and food safety, research and experimental development has already given concrete proof of its potential as an indispensable partner in the resilience and recuperation of soils for agricultural production.

8. Conclusions

The role of insects in the biological digestion of organic substrates, with a view to the fertilizing potential of entomocomposts, has raised a growing commitment from the scientific and technical community in the field of agriculture and environmental protection; however, despite the accumulation of positive results from the application of this type of fertilizer, significant progress is still expected in this sector, with the improvement of the genetic capacity of insects, of the pre-treatment of substrates and of the entomocomposting technology, so as the adequacy of fertilization techniques.

In the context of the organic fertilizers, entomocomposting takes precedence over other composting methods, mainly because of the speed of the organic waste digestion process, drastically reducing composting time and thus the risk of environmental pollution, besides advantages such as soil health, pest control, sprouting and germination potential.

Various insects have been tested for their potential in digesting substrates of a very different nature, giving rise to entomocomposts with positive results, in reduce mineral fertilizers, in crop production, or as correctors of certain chemical and/or microbiological deficiencies, not to mention physical soil deficiencies, for which any organic fertilizer is capable of dealing with.

Nevertheless, more results are expected with further research into entomocomposting technology, with the discovery of new insect species and their genetic improvement for the biodigestion of different organic substrates, and with new techniques for the enhancement of the fertilizing effect of composts, in order to make available suitable formulations for different "soil x plant x fertilizer" interaction situations.

Until now, as shown in the tests presented in this analysis, the greatest success in the contentious debate "organic vs. mineral" has been achieved in situations of compromise, where the organic fertilizer has the complementary role, by its relatively low and unbalanced nutrient content, notwithstanding its biological interaction with plant and soil microrganisms, its action in improving the soil's physical properties and its capacity to retain water and nutrients – so, as advocated by Ronald and Adamchak [62] or Amman K. [63], and as Saint Tomas d'Aquino said so well, '*in medio stat virtus*'.

Furthermore, although growing exponentially, increased production of organic waste for entomocomposting is unlikely to be sufficient to ensure global food security on its own, as it is a direct function of population growth; suggestions based on success rates reported in the literature for insect frass - ranging from 10 to 40% by volume - may be realistic to be expected, at least in the medium term. In fact, if the potential of the triple valence of entomocomposting (protection of the environment, food security and resilient soil fertility) can already be categorically stated, the use of entomocompost as a fertilizer still faces the major constraint of the lack of scale of its production.

Considering all these possibilities, insects must be recognized not only as a nutrient source but also as a tool. The value of insects can surpass the production of nutrients and the use of its by/co-products to increase its profitability. In the near future insects could be used in manure and household waste treatment approaches, decreasing the environmental impact of livestock production and landfill volumes [5, 64, 65]. This approach would open a completely new opportunity for insect rearing, that is distinct from insect production for animal nutrition which must comply with safety and hygiene regulations.

Increased sustainability of animal and food production can be delivered by insect use, not only through the development of new feed resources but also by contributing to the reduction and conversion of wastes into novel raw materials for bioindustry and biorefinery approaches.

There is still much to do in this regard but, in rural areas, the proposal of a circular economy system in the management of agricultural, livestock and forestry production, with circularities within private farms to be extrapolated (cooperatively) to the regional level with agroindustry and an industrial entomocomposting unit, deserved to be weighed up.

Acknowledgements

This chapter was performed under the scope of the NETA project: New Strategies in Wastewater Treatment (POCI-01-0247- FEDER-046959) funded by PORTUGAL2020.

Author details

Regina Menino[1] and Daniel Murta[2,3,4*]

1 Instituto Nacional de Investigação Agrária e Veterinária, IP (INIAV), Oeiras, Portugal

2 EntoGreen-Ingredient Odyssey SA, Santarém, Portugal

3 CiiEM-Centro de investigação interdisciplinar Egas Moniz, Caparica, Portugal

4 Myrtus Unipessoal Lda, Nisa, Portugal

*Address all correspondence to: daniel.murta@entogreen.com

References

[1] Čičková H., Newton G., Lacy R. & Kozánek M. 2015. The use of fly larvae for organic waste treatment. *Waste Manag.*, **35**:68-80

[2] Kummu M., de Moel H., Porkka M., Siebert S., Varis O. & Ward P.J. 2012. Lost food, wasted resources: Global food supply chain losses and their impacts on freshwater, cropland, and fertiliser use. *Sci. Tot. Environm.*, **438**:477-489. doi.org/10.1016/j.scitotenv.2012.08.092

[3] Barroso F.G., de Haro C., Sánchez-Muros M.J., Venegas E., Martínez-Sánchez A. & Pérez-Bañón C. 2013. The potential of various insect species for use as food for fish. *Aquaculture*, 422-423, 193-201. doi: 10.1016/j.aquaculture.2013.12.024

[4] Tran G., Heuzé V. & Makkar H.P.S. 2015. Insects in fish diets. *Animal Frontiers*, **5(2)**:37-44. doi.org/10.2527/af.2015-0018

[5] van Raamsdonk L.W.D., van der Fels-Klerx H.J. & de Jong J. 2017. New feed ingredients: the insect opportunity. *Food Additives and Contaminants - Part A*, **34(8)**:1384-1397. doi.org/10.1080/19440049.2017.1306883

[6] Veldkamp T. & Bosch G. 2015. Insects : a protein-rich feed ingredient in pig and poultry diets. *Animal Frontiers*, **5(2)**:45-50. doi.org/10.2527/af.2015-0019

[7] Sprangers T., Ottoboni M., Klootwijk C., Ovyn A., Deboosere S., De Meulenaer B., Michiels J., Eeckhout M., De Clercq P. & De Smet S. 2017. Nutritional composition of black soldier fly *(Hermetia illucens)* prepupae reared on different organic waste substrates. *J. Sci. Food and Agric.*, **97**:2594-2600. doi.org/10.1002/jsfa.8081

[8] Pinotti L., Giromini C., Ottoboni M., Tretola M. & Marchis D. 2019. Review: Insects and former foodstuffs for upgrading food waste biomasses/streams to feed ingredients for farm animals. *Animal*, **13(7)**:1365-1375. doi.org/10.1017/S1751731118003622

[9] de Bertoldi M., Vallini G. & Pera A. 1983. The biology of composting: A review. *Waste Management & Research*, **1(2)**:157-176. doi:10.1016/0734-242x(83)90055-1

[10] Gusmão L., Mexia J.T. & Gomes M.L. 1988. Mapping of equipotential zones for cultivar yield patterns evaluation. *Plant Breeding*, **103**:293-298

[11] Gusmão L., Mexia J.T. & Baeta J. 1992. Trimmed Joint Regression: A new approach to the Joint Regression Analysis for cultivar relative performance evaluation. *Theor. Appl. Genet.*, 84:735-738

[12] Chen J.H. 2006. The combined use of chemical and organic fertilizers and/or biofertilizer for crop growth and soil fertility. *Int. Workshop Sustained Manag. Soil-Rhizosphere Syst. for Efficient Crop Prod. and Fertilizer Use*. Bangkok, 1-11

[13] Zhang Z., Elser J.J., Cease A.J., Zhang X., Yu Q., Han X. & Zhang G. 2014. Grasshoppers regulate N:P stoichiometric homeostasis by changing phosphorus contents in their frass. *PloS One*, **9**:e103697. doi.org/10.13 71/journal.pone.0103697

[14] Menino R. & Murta D. 2021. BSF - time to change the flies. *Hort. Int. J.*, **5(3)**:114-117. doi: 10.15406/hij.2021.05.00215

[15] Timsina J. 2018. Can organic sources of nutrients increase crop yield to meet global food demand? *Agronomy*, **8**:214; doi:10.3390/agronomy8100214

[16] Poveda J., Jiménez-Gómez A., Saati-Santamaría Z., Usategui-

Martín R., Rivas R. & García-Fraile P. 2019. Mealworm frass as a potential biofertilizer and abiotic stress tolerance-inductor in plants. *Applied Soil Ecol.*, **142**:110-122. doi:10.1016/j.apsoil.2019.04.016

[17] Koh D.W.S., Ang B.Y.X., Yeo J.Y., Xing Z. & Gan S.K. 2020. Plastic agriculture using worms: Augmenting polystyrene consumption and using frass for plant growth towards a zero-waste circular economy. bioRxiv. doi.org/10.1101/2020.05.29.123521

[18] Zhu F.X., Yao Y.I., Wang S.J., Du R.G., Wang W.P., Chen X.Y., Hong C.L., Qi B., Xue Z.Y. & Yang H.Q., 2015. Housefly maggot-treated composting as sustainable option for pig manure management. *Waste Manag.*, **35**:62-67. doi.org/10.1016/j.wasman.2014.10.005

[19] Matthäus B., Piofczyk T., Katz H. & Pudel F. 2019. Renewable Resources from Insects: Exploitation, Properties, and Refining of Fat Obtained by Cold-Pressing from Hermetia illucens (Black Soldier Fly) Larvae. *Europ. J. Lipid Sci. Technol.*, **121(7)**:1800376. doi.org/10.1002/ejlt.201800376

[20] Song Y.-S.; Kim M.-W.; Moon C.; Seo D.-J.; Han Y.S.; Jo Y.H.; Noh M.Y.; Park Y.-K.; Kim S.-A.; Kim Y.W.; Jung W.-J. 2018. Extraction of chitin and chitosan from larval exuvium and whole body of edible mealworm, *Tenebrio molitor. Entomological Research*, **48**(3):227-233. doi: 10.1111/1748-5967.12304

[21] Aider M. 2010. Chitosan application for active bio-based films production and potential in the food industry: Review. *LWT - Food Sci. Technol.*, **43(6)**:837-842

[22] Kong M., Chen X.G., Xing K. & Park H.J. 2010. Antimicrobial properties of chitosan and mode of action: A state of the art review. *Int. J. Food Microbiol.*, **144(1)**:51-63

[23] Verlee A., MinCke S. & Stevens C.V. 2017. Recent developments in antibacterial and antifungal chitosan and its derivatives. *Carbohydrate Polymers*, **164**:268-283. doi.org/10.1016/j.carbpol.2017.02.001

[24] Sharp R. 2013. A Review of the Applications of Chitin and Its Derivatives in Agriculture to Modify Plant-Microbial Interactions and Improve Crop Yields. *Agronomy*, **3(4)**:757-793. https://doi.org/10.3390/agronomy3040757

[25] Houben D., Daoulas G., Faucon M.P. & Dulaurent A.M. 2020. Potential use of mealworm frass as a fertilizer: Impact on crop growth and soil properties. Scientific Reports, **10(1)**. doi:10.1038/s41598-020-61765-x

[26] Poveda, J. 2021. Insect frass in the development of sustainable agriculture. A review. *Agron. Sustain. Dev.*, **41**:5. doi.org/10.1007/s13593-020-00656-x

[27] Lardé G. 1990. Recycling of coffee pulp by *Hermetia illucens* (Diptera: *Stratiomyidae*) larvae. *Biological Wastes*, **33(4)**:307-310

[28] Smetana S., Palanisamy M., Mathys A. & Heinz V. 2016. Sustainability of insect use for feed and food: Life cycle assessment perspective. *J. Cleaner Prod.*, **137**:741-751. doi.org/10.1016/j.jclepro.2016.07.148

[29] Fowles T.M. & Nansen C. 2020. Insect-Based Bioconversion: Value from Food Waste. *In*: Närvänen E., Mesiranta N., Mattila M., Heikkinen A. (eds). Food Waste Management. Palgrave Macmillan, Cham. doi.org/10.1007/978-3-030-20561-4_12

[30] James M.T. 1935. The genus *Hermetia* in the United States (Diptera:

Stratiomyidae). Bull. Brooklyn Entomol. Soc., **30**:165-170

[31] Sheppard D.C., Newton G.L., Thompson S.A. & Savage S. 1994. A value-added manure management-system using the black soldier fly. *Bioresour. Technol.*, **50**:275-279. doi.org/10.1016/0960-8524(94)90102-3

[32] Singh A. & Kumari K. 2019. An inclusive approach for organic waste treatment and valorisation using Black Soldier Fly larvae: A review. *J. Environm. Manag.*, **251**:109569. doi:10.1016/j.jenvman.2019.109569

[33] Wang Y.S. & Shelomi M. 2017. Review of black soldier fly (*Hermetia illucens*) as animal feed and human food. *Foods, 6*(**10**):91. doi:10.3390/foods6100091

[34] Dulaurent A.M., Daoulas G., Faucon M.P. & Houben D. 2020. Earthworms (*Lumbricus terrestris* L.) mediate the fertilizing effect of frass. *Agronomy*, **10**:783. https://doi.org/10.3390/agronomy10060783

[35] Zink T. & Geyer R. 2017. "Circular Economy Rebound". *J. Ind. Ecol.*, **21(3)**:593-602. doi:10.1111/jiec.12545. S2CID 157110158

[36] Allwood J.M. 2014. "Squaring the Circular Economy". *Handbook of Recycling*, 445-477. doi:10.1016/b978-0-12-396459-5.00030-1. ISBN 978-0-12-396459-5

[37] Yildirim-Aksoy M., Eljack R. & Beck B.H. 2020a. Nutritional value of frass from black soldier fly larvae, *Hermetia illucens*, in a channel catfish, *Ictalurus punctatus*, diet. *Aquacult Nutr*, **26**:812-819. doi.org/10.1111/anu.13040

[38] Yildirim-Aksoy M., Eljack R., Schrimsher C. & Beck B.H. 2020b. Use of dietary frass from black soldier fly larvae, *Hermetia illucens*, in hybrid tilapia (Nile x Mozambique, *Oreochromis niloticus* x *O. Mozambique*) diets improves growth and resistance to bacterial diseases. *Aquacult Rep*, **17**:100373, doi.org/10.1016/j.aqrep.2020.100373

[39] Alattar M.A., Alattar F.N. & Popa, R. 2016. Effects of microaerobic fermentation and black soldier fly larvae food scrap processing residues on the growth of corn plants (*Zea mays*). *Plant Sci. Today*, **3(1)**:57-62. dx.doi.org/10.14719/pst.2016.3.1.179

[40] Gärttling D., Kirchner S.M. & Schulz H. 2020. Assessment of the N- and P-Fertilization Effect of Black Soldier Fly (Diptera: Stratiomyidae) By-Products on Maize, *J. Insect Sci.*, **20(5)**:1-11, doi.org/10.1093/jisesa/ieaa089

[41] Houben D., Daoulas G. & Dulaurent A.M. 2021. Assessment of the short-Term Fertilizer Potential of Mealworm Frass Using a Pot Experiment. *Front. Sust. Food Syst.*, **5**:714596. doi:10.3389/fsufs.2021.714596

[42] Newton G.L., Sheppard D.C., Watson D.W., Burtle G. & Dove R. 2005. Using the Black Soldier Fly, *Hermetia Illucens*, as a Value-Added Tool for the Management of Swine Manure. Animal and Poultry Waste Management Center, North Carolina State University, Raleigh, NC. http://www.organicvaluerecovery.com/studies/studies_htm_files/bsf_value_added.pdf

[43] Choi Y., Choi J., Kim J., Kim M., Kim W., Park K., Bae S. & Jeong G. 2009. Potential Usage of Food Waste as a Natural Fertilizer after Digestion by *Hermetia illucens* (Diptera: Stratiomyidae). *Int. J. Industr. Entomol., 19*:171-174.

[44] Anggraeni D. 2010. The effect of Bioconversion Fertilizer Palm Kernel Meal (BFPKM) as fertilizer for the growth of *Vigna unguiculata* L. Walp (yardlong bean) var. mutiara. Thesis for

S2 graduation. University of Indonesia, 95pp

[45] Kebli H. & Sinaj S. 2017. Potential agronomique d'un engrais naturel a base de digestats de larves de mouches. Recherche *Agronomique Suisse*, **8(3)**:88-95

[46] Esteves, C.F.M. 2020. Utilização do composto orgânico de larvas da *Hermitia illucens* como fertilizante em alface. Dissertação para obtenção de grau de Mestre. Instituto Superior de Agronomia (ISA), Universidade de Lisboa. 88pp.

[47] Klammsteiner T., Turan V., Fernández-Delgado Juárez M., Oberegger S. & Insam H. 2020. Suitability of Black Soldier Fly Frass as Soil Amendment and Implication for Organic Waste Hygienization. *Agronomy, 10*:1578. doi.org/10.3390/agronomy10101578

[48] Beesigamukama D., Mochoge B., Korir N.K., Fiaboe K.K.M., Nakimbugwe D., Khamis F.M., Subramanian S., Dubois T., Musyoka M.W., Ekesi S., Kelemu S. & Tanga C.M. 2020. Exploring Black Soldier Fly Frass as Novel Fertilizer for Improved Growth, Yield, and Nitrogen Use Efficiency of Maize Under Field Conditions. Front. Plant Sci. **11**:574592. doi: 10.3389/fpls.2020.574592

[49] Chirere T.E.S., Khalil S. & Lalander C. 2021. Fertiliser effect on Swiss chard of black soldier fly larvae-frass compost made from food waste and faeces. *J. Insects as Food and Feed*, **7(4)**:457-469. doi.org/10.3920/JIFF2020.0120

[50] Rosmiati M., Nurjanah K.A., Suantika G. & Putra R.E. 2017. Application of Compost Produced by Bioconversion of Coffee Husk by Black Soldier Fly Larvae (*Hermetia Illucens*) as Solid Fertilizer to Lettuce (*Lactuca Sativa* Var. Crispa): Impact to Growth. *Proc. Int. Conf. Green Technol.*, **8(1)**:38-44

[51] Setti L., Francia E., Pulvirenti A., Gigliano S., Zaccardelli M., Pane C., Caradonia F., Bortolini S., Maistrello L. & Ronga D. 2019. Use of black soldier fly (*Hermetia illucens* (L.), Diptera: Stratiomyidae) larvae processing residue in peat-based growing media. *Waste Manag.*, **95**:278-288. doi:10.1016/j.wasman.2019.06.017

[52] Kawasaki K., Kawasaki T., Hirayasu H., Matsumoto Y. and Fujitani Y. 2020. Evaluation of Fertilizer Value of Residues Obtained after Processing Household Organic Waste with Black Soldier Fly Larvae (*Hermetia illucens*). *Sustainability*, **12(12)**:4920. doi.org/10.3390/su12124920

[53] Temple W.D., Radley R., Baker-French J. & Richardson F. 2013. *Use of Enterra natural fertilizer (black soldier fly larvae digestate) as a soil amendment*. Research Final Report, Enterra Feed Corporation, Vancouver, Canada. 34pp

[54] Anyega A.O., Korir N.K., Beesigamukama D., Changeh G.J., Nkoba K., Subramanian S., van Loon J.J.A., Dicke M. & Tanga C.M. 2021. Black Soldier Fly-Composted Organic Fertilizer Enhances Growth, Yield, and Nutrient Quality of Three Key Vegetable Crops in Sub-Saharan Africa. *Front. Plant Sci.*, **12**:680312. doi: 10.3389/fpls.2021.680312

[55] Quilliam R.S., Nuku-Adeku C., Maquart P., Little D., Newton R. & Murray F. 2020. Integrating insect frass biofertilisers into sustainable peri-urban agro-food systems. *J. Insects as Food and Feed*, **6(3)**:315-322. doi:10.3920/jiff2019.0049

[56] Matos S. & Murta D. 2019. Projeto ENTOVALOR e a valorização de subprodutos. *Revista Agrotejo*, **29**:

106-108. https://agrotejo.repo.pt/revista-agrotejo-29/

[57] Putra R.E., Hutami R., Suantika G. & Rosmiati M. 2017. Application of compost produced by bioconversion of coffee husk by Black Soldier Fly Larvae (*Hermetia Illucens*) as solid fertilizer to lettuce (*Lactuca sativa* Var. Crispa): impact to harvested biomass and utilization of nitrogen, phosphor, and potassium. Proc. Int. Conf. Green Technol., **8(1)**:466-472

[58] Choi S. & Hassanzadeh N. 2019. BSFL Frass: A Novel Biofertilizer for Improving Plant Health While Minimizing Environmental Impact. *Candian Sci. Fair J.*, **2**:41-46. doi: 10.18192/csfj.v2i220194146

[59] Elissen H., Schilder M., Postma J., van der Weide R. 2019. Disease suppression in cress and sugar beet seedlings with frass of the black soldier fly (*Hermetia illucens*). Stichting Wageningen Research, Wageningen Plant Research, Business Unit Field Crops

[60] Kagata H. & Ohgushi T. 2011. Positive and negative impacts of insect frass quality on soil nitrogen availability and plant growth. *Pop. Ecol.,* **54**(1):75-82. doi: 10.1007/S10144-011-0281-6

[61] Menino R., Felizes F., Castelo-Branco M.A., Fareleira P., Moreira O., Nunes R. & Murta D. 2021. Agricultural value of Black Soldier Fly larvae frass as organic fertilizer on ryegrass. *Heliyon*, Jan 2, **7(1)**:e05855. doi: 10.1016/j.heliyon.2020.e05855. PMID: 33426352, PMCID: PMC7785954

[62] Ronald P.C. & Adamchak R.W. 2008. *Tomorrow's Table: Organic Farming, Genetics and the Future of food*; Oxford University Press: NY, USA.

[63] Amman K. 2009. Why farming with high tech methods should integrate elements of organic agriculture. *New Biotech.*, **25**:378-388

[64] Li Q., Zheng L., Qiu N., Cai H., Tomberlin J.K. & Yu Z. 2011. Bioconversion of dairy manure by black soldier fly (Diptera: Stratiomyidae) for biodiesel and sugar production. *Waste Manag.*, **31(6)**:1316-1320. doi.org/10.1016/j.wasman.2011.01.005

[65] Surendra K.C., Olivier R., Tomberlin J.K., Jha R. & Khanal S.K. 2016. Bioconversion of organic wastes into biodiesel and animal feed via insect farming. *Renewable Energy*, **98**:197-202. doi.org/10.1016/j.renene.2016.03.022